Neural Networks and Computing

Learning Algorithms and Applications

SERIES IN ELECTRICAL AND COMPUTER ENGINEERING

Editor: Wai-Kai Chen *(University of Illinois, Chicago, USA)*

Published:

Vol. 1: Net Theory and Its Applications
Flows in Networks
by W. K. Chen

Vol. 2: A Mathematical Introduction to Control Theory
by S. Engelberg

Vol. 5: Security Modeling and Analysis of Mobile Agent Systems
by Lu Ma and Jeffrey J. P. Tsai

Vol. 6: Multi-Objective Group Decision Making: Methods Software and Applications with Fuzzy Set Techniques
by Jie Lu, Guangquan Zhang, Da Ruan and Fengjie Wu

SERIES IN ELECTRICAL AND COMPUTER ENGINEERING VOL. 7

Neural Networks and Computing

Learning Algorithms and Applications

Tommy W S Chow

City University of Hong Kong, Hong Kong

Siu-Yeung Cho

Nanyang Technological University, Singapore

ICP Imperial College Press

Published by

Imperial College Press
57 Shelton Street
Covent Garden
London WC2H 9HE

Distributed by

World Scientific Publishing Co. Pte. Ltd.
5 Toh Tuck Link, Singapore 596224
USA office: 27 Warren Street, Suite 401-402, Hackensack, NJ 07601
UK office: 57 Shelton Street, Covent Garden, London WC2H 9HE

British Library Cataloguing-in-Publication Data
A catalogue record for this book is available from the British Library.

NEURAL NETWORKS AND COMPUTING
Series in Electrical and Computer Engineering — Vol. 7

ISBN-13 978-1-86094-758-2
ISBN-10 1-86094-758-1

Printed in Singapore by World Scientific Printers (S) Pte Ltd

Preface

The area of Neural computing that we shall discuss in this book represents a combination of techniques of classical optimization, statistics, and information theory. Neural network was once widely called artificial neural networks, which represented how the emerging technology was related to artificial intelligence. It once was a topic that had captivated the interest of most computer scientists, engineers, and mathematicians. Its charm of being an adaptive system or universal functional approximator has compelling appeal to most researchers and engineers. The Backpropagation training algorithm was once the most popular keywords used in most engineering conferences. There is an interesting history on this area dated back from the late fifties which we saw the advent of the Mark I Perceptron. But the real intriguing history started from the sixties that we saw Minsky and Papert's book "*Perceptrons*" discredited the early neural research work. For all neural researchers, the late eighties are well remembered because the research of neural networks was reinstated and repositioned. From the nineties to the new millennium is history to be made by all neural researchers. We saw the flourish of this topic and its applications stretched from rigorous mathematical proof to different physical science and even business applications. Researchers now tend to use the term "neural networks" instead of "artificial neural networks" when we have understood the theoretical background more. There have been volumes of research literature published on the new development of neural theory and applications. There have been many attempts to discuss this topic from either a very mathematical way or a very practical way. But to most users including students and engineers, how to employ an appropriate neural network learning algorithm and the selection of model for a given physical problem appear to be the main issue.

This book, written from a more application perspective, provides thorough discussions on neural network learning algorithms and their related issues. We strive to find the balance in covering the major topics in neurocomputing, from learning theory, learning algorithms, network architecture to applications. We start the book from the fundamental building block "neuron" and the earliest neural network model, McCulloh and Pitts Model. In the beginning, we treat the learning concept from the well-known regression problem which shows how the idea of data fitting can be used to explain the fundamental concept of neural learning. We employ an error convex surface to illustrate the optimization concept of learning algorithm. This is important as it shows readers that the neural learning algorithm is nothing more than a high dimensional optimization problem. One of the beauties of neural network is being a soft computing approach in which the selection of a model structure and initial settings may not have noticeable effect on the final solution. But neural learning process also suffers from its problem of being slow and stuck in local minima especially when it is required to handle a rather complex problem. These are the two main issues addressed in the later chapters of this book. We study the neural learning problem from a new perspective and offer several modified algorithms to enhance the learning speed and its convergence ability. We also show initializations of a network have significant effect on the learning performance. Different initialization methods are then discussed and elaborated.

Later chapters of the book deal with Basis function network, Self-Organizing map, and Feature Selection. These are interesting and useful topics to most engineering and science researchers. The Self-Organizing map is the most widely used unsupervised neural network. It is useful for performing clustering, dimensional reduction, and classification. The SOM is very different from the feedforward neural network in the sense of architecture and learning algorithm. In this book, we have provided thorough discussions and newly developed extended algorithms for readers to use. Classification and Feature selection is discussed in Chapter 6. We include this topic in the book because bioinformatics has recently become a very important research area. Gene selection using computational methods, and performing cancer classification computationally have become the 21st Century research. This book provides a detail discussion on feature selection and how different methods be applied to gene selection and cancer classification. We hope this book will provide useful and inspiring information to readers. A number of software algorithms written in MATLAB are available for readers to use. Although the authors have gone through the book for few times checking typos and errors, we would appreciate readers notifying us about any typos found.

At last, the authors would thank the support and help from his colleagues and students. The author must thank his students Mr. Gaoyang Dai, Dr. Di Huang, Dr. Jim Y F Yam, Dr. Sitao Wu, and Dr. M K M Rahman for their help in preparing the material. The authors would also thank Prof. Guanrong Chen for introducing us to the very helpful publisher. The authors must thank the World Scientific Publishing Co for publishing this book.

Last but not least, Tommy Chow would like to thank Irene, Adrian and Ian for their tolerance and support.

Contents

Chapter 1

Introduction

1.1. Background

Human beings are now living in an era of unparalleled changes especially in the world of science and technology. We have witnessed the impact of DNA research from the way of diagnosing cancer, creating new drugs and to the way of tracing our human ancestors back to over eighty thousand years ago. We have also witnessed the birth of the Information Technology era which has changed our life from the personal entertainment habits to our learning and even the way of how our business is run. All these amazing things, which are attributed to the immense development of computing technology, happened in merely few decades. Undoubtedly, computers have become an integral part of our lives that most users have taken them for granted but may not know that the structure of computer is based on John Von Neumann's view on how computational process be organized. Von Neumann architecture is a very organized way of processing computation. The computing is achieved by using a CPU operating a series of instructions including: fetch, decode, execute, and writeback. This type of serial operation has been working extremely well for virtually all applications, until the challenge posed by the emerging Artificial Intelligent technology.

Neural Computing is basically a parallel distributed processing. It has the ability of performing supervised and/or unsupervised learning to adapt the information environment. The architecture of neural network is in fact based on the way of how our human nerve system is connected. Generally, there are about 100 billion numbers of neuron in a human brain. Neurons in our brain are parallel connected to numerous other neurons forming a massive parallel computer like machine. Neural network is designed in a way to seek the style of computing of human brain. As a result, neural network is powerful enough to solve a variety of problems that are proved to be difficult with conventional digital computational methods. Typical cognitive tasks include recognition of a familiar face, learning to speak and understand a natural language, retrieving contextually appropriate information from memory, and performing demanding classification tasks.

The human thinking system is in parallel which means it operates with numerous of our neurons connected together. In contrast to the conventional crisp mathematical logic, the main characteristics of human thinking process is imprecise, fuzziness, but adaptive. It learns by examples, experience and it exhibits strong adaptation to external changes. Neural networks are designed in a way to mimic most of these characteristics. They are so far exhibiting very encouraging performance.

Learning: Neural network can modify their behaviour in response to the environment. When a given set of inputs with or without desired outputs, they can self-adjust to produce consistent responses.

Generalization: When the network is once trained, its response can be in some extent insensitive to minor variations, which may be caused by noise corruption or slight distortion in a real-world environment, in its inputs.

Massively parallelism: Information is processed in a massive parallel manner.

Fault-tolerance: Once when the network connections are made, the network is able to deliver a robust behaviour. The response of the network as a whole may only be slightly degraded if some of its processing elements are slightly damaged or altered.

1.2. Neuron Model

Neuron is the fundamental cellular unit of a nervous system. A typical biological neuron in human brain is shown in Fig. 1.1. In each neuron it has an output fiber called axon, and a button like terminal called synapse. The axon, which is the output path of a neuron, splits up and connects to many dendrites, which are the input paths of other neurons, through a junction called synapse. Each neuron receives and combines signals from numerous neurons through dendrites similarly connected.

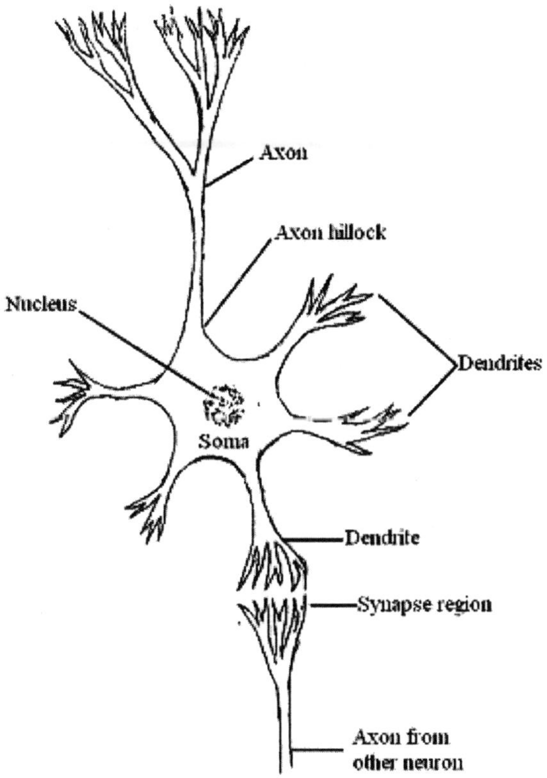

Figure 1.1. A typical biological neuron and its connection

A neuron can receive up to few thousands to about fifteen thousands inputs from the axons of other neurons. Apparently this forms a massive parallel system compared with the digital computer architecture. In each neuron, if the combined signal exceeds a threshold value, it activates the firing of a neuron producing an output signal through the axon. Transmission of information across synapses is in fact a chemical in nature, but it has electrical side effects which we can be measured. Electrical activities in neurons appeared in a shape of pulse with a frequency in the range of 1 KHz. This type of biological behaviour is modeled by an electronic model shown in Fig. 1.2. A simple neuron model is the most fundamental processing element of neural networks. The weights correspond to the synaptic strength of neural connections, i.e., analogous to "memory" and the neuron sums all the weighted inputs, modifies the signals through a transfer function which is usually non-linear. The transfer function can be a threshold function which only allows signal to be passed if the combined activity level reaches a certain threshold level, or it is a continuous function of the combined input.

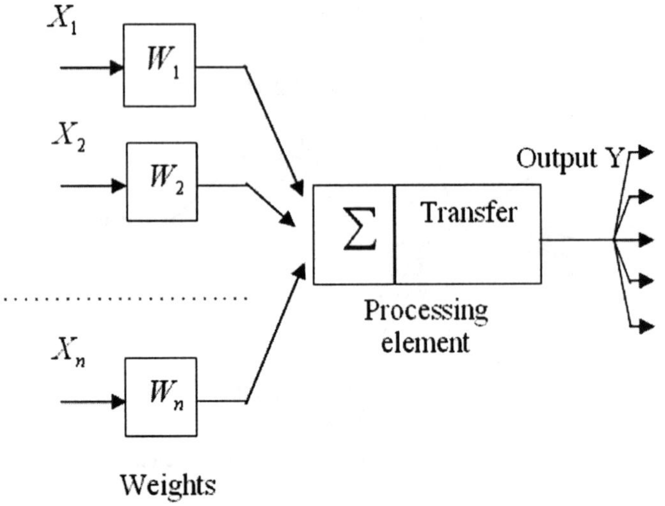

Figure 1.2. Electronic model of a neuron

1.3. Historical Remarks

The first neural network model has to be the McCulloch and Pitts model proposed in 1943. It is a very simple electronic model that can be hardware implemented. It is a multiple inputs summing device that consists of different weightings for each input, and a threshold before the output. The significance of this model at that time was its ability to compute any arithmetic or logical function. It is until about the ends of the fifties that the first type of Perceptron was proposed by Frank Rosenblatt, and Wightman. The first application of the Mark I Perceptron machine was on pattern recognition. In their experiments, the pattern recognition ability of Perceptron model was demonstrated by recognizing different simple characters. Subsequently, numerous neural networks results were reported. Neural hardware business was even established. It was similar to the hype of the IT industries in the end of nineties, the neural hype burst when good idea ran out. By the mid of sixties, Minsky and Papert's paper mathematically proved that a simple perceptron could not handle the XOR function. This has silenced neural research work for about 2 decades. It was until about the mid of eighties that Neural research has become flourish again. It was achieved with the contribution from a number of renowned scientists including Hopfield, Amari, Grossberg, Kohonen, Fukushima, Oji, etc. They have developed many important neural topologies.

In 1982, Hopfield published two important articles which most people still regard as the inauguration of the current neural network era. Hopfield showed a novel idea that models of physical systems could be effectively used for solving computational problems. Using the idea of an energy function to establish a new type of network architecture, he showed that when a distorted pattern is presented to the network, the pattern is associated with another pattern which belongs to a similar class pattern. Hopfield networks are, thus, sometimes called associative networks. For a discrete-time Hopfield network, the energy of a certain vector with a given initial state will converge to a value having minimum energy. This is used to explain its capability of converging to a similar class of patterns. Also, in the early eighties Carpenter and Grossberg established the well-known adaptive resonance theory (ART) based on their early work on competitive learning. ART was introduced by them over the period of 1976-1986 as a theory of human information processing. Like Kohonen's Self Organising Map, they were working on systems that are capable of organizing themselves. ART has the ability to learn without supervised training and is consistent with cognitive and behavioral models. It was derived based on competitive learning which is capable of finding categories. ART has been widely used as a type of neural network models that perform supervised and unsupervised category learning, and pattern classification.

In 1988 George Cybenko published a very important work proving the universal functional approximation ability of neural networks. In 1989, Funahashi, Hornik, and Stinchcombe also reported their findings on proving multilayer Perceptron network as a universal approximator. Subsequently, neural networks have been widely applied on many different science and engineering areas. The nowadays neural networks have already extended from its simple pattern recognition problem to the very complicated DNA, gene recognition and classification problems. Applications have extended from physical science and engineering to finance, economic and social science.

1.4. Network architecture

Generally, neural networks can be categorized into 2 main types, namely supervised networks and unsupervised networks. The way the network architecture was deigned has taken the ability of its training algorithm into account. In most newly proposed network topologies, the design of their corresponding training algorithms are deemed essential. Apparently, a successful network architecture must be supported by an effective and simple enough training algorithm. In this book, the later chapter will detail different efficient training algorithms. In this section, we focus the introduction of their hardware architectures.

1.4.1. Supervised Neural Networks

Supervised neural networks are the mainstream of neural network development. The differentiable characteristic of supervised neural networks lies in the inclusion of a teacher in their learning process. The basic block diagram of supervised learning for all neural network models can be described through Fig. 1.3. For the learning process, the network needs training data examples consisting of a number of input-output pairs. The desired output vector in the training data set serve as a teacher for the network's learning. In the training process the error signals are constructed from the difference between the desired output and system output. Through iterative training procedure the network's weights are adjusted by the error signal in a way that the network output tries to follow the desired output as close as possible.

The learning procedure continues until the error signal is close to zero or below a predefined value. The sum of errors over all the training samples can be considered as a kind of network performance measure, which is a function of free parameters of the system. Such function can be visualized a multidimensional error surface where network free parameters serves as coordinates. During the course of learning the system gradually moves to a minimum point along an error surface. The error surface is determined by the network architecture and the cost function. In later chapter of this book, more details in this aspect will be discussed. The supervised networks' architectures can vary depending on the complexity and nature of data to be handled. Broadly speaking, they can be sub-divided into following three fundamental classes.

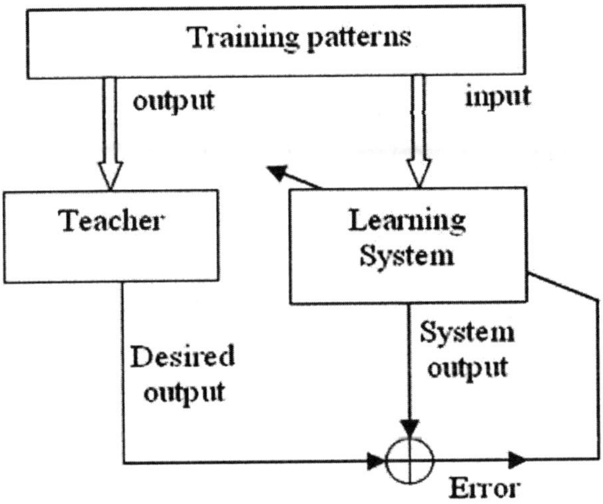

Figure 1.3. Overview of supervised learning

1.4.1.1. McCulloh and Pitts Model

In 1943, neurophysiologist McCulloch and his associate Pitts used a type of electrical circuit to model a simple neuron. This is the first ever type of electronic model used to model a neuron. The McCulloh and Pitts (MCP) model has a profound impact on the later Perceptron model. The MCP model basically consists of a summing amplifier and variable resistors as shown in Fig. 1.4. The weights (W_1, W_2) in Fig. 1.4 are adjusted through varying the values of the resistors. A threshold device, which is made of voltage comparator, generates an output 1 if the summation of signals exceeds the threshold value, T, or the output is 0 if the signal is less than the threshold value, T.

To illustrate how the MCP model works, we use a simple "AND" logic problem as example.The general form of MCP model is

$$Output = 1, \text{ if } X_1W_1 + X_2W_2 > T \quad (1.1)$$

We make use of the truth table and the following inequalities are obtained.

$$O < T\,,\ O + W_2 < T\,,\ W_1 + O < T\,,\ W_1 + W_2 > T$$

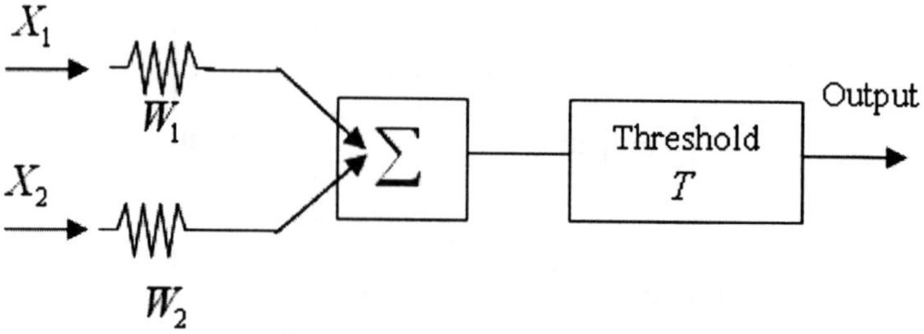

Figure 1.4. The general form of MCP model

Table 1.1

X_1	X_2	Output
0	0	0
0	1	0
1	0	0
1	1	1

The task is to determine the values of T, W_2 and W_2 so that the output satisfies the logic "AND" function. We can chose $W_1 = 0.5$, $W_2 = 0.8$ and $T = 1$ as a solution. Fig. 1.5 shows that the job is equivalent of finding a line separating the 3 "0" from the "1". Apparently, the 3 lines and many other lines can satisfy the requirement.

To find the separating line, we need to find the gradient and the interceptions of the line. By replacing ">" with "=",

$$W_1X_1 + W_2X_2 = T \tag{1.2}$$

$$X_2 = \frac{T}{W_2} - \frac{W_1X_1}{W_2}$$

Thus $a = \frac{1}{0.8} = 1.25$, and $gradient = \frac{-5}{8}$, $b = \frac{1}{0.5} = 2$

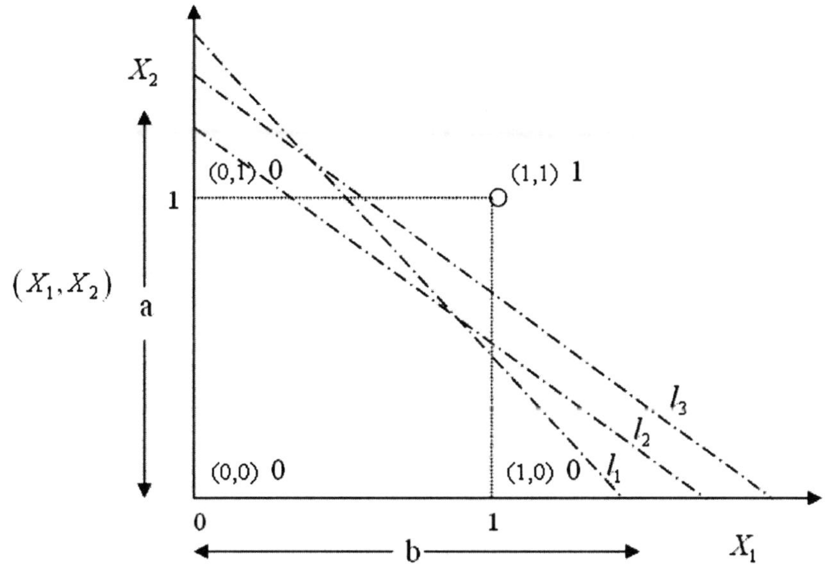

Figure 1.5.

There are infinite numbers of solutions that can satisfy the requirement. Clearly, it is a very lose way in finding a solution which appears to be one of the beauty of neural network. The Application of a "AND" logic function may appear to be too simple to illustrate its concept of finding a hyper-plane for discrimination. We now consider the following logic function.

Table 1.2

X_1	X_2	X_3	Output
0	0	0	1
0	1	1	0
0	0	1	0
0	1	0	1
1	0	0	0
1	1	1	0
1	0	1	0
1	1	0	0

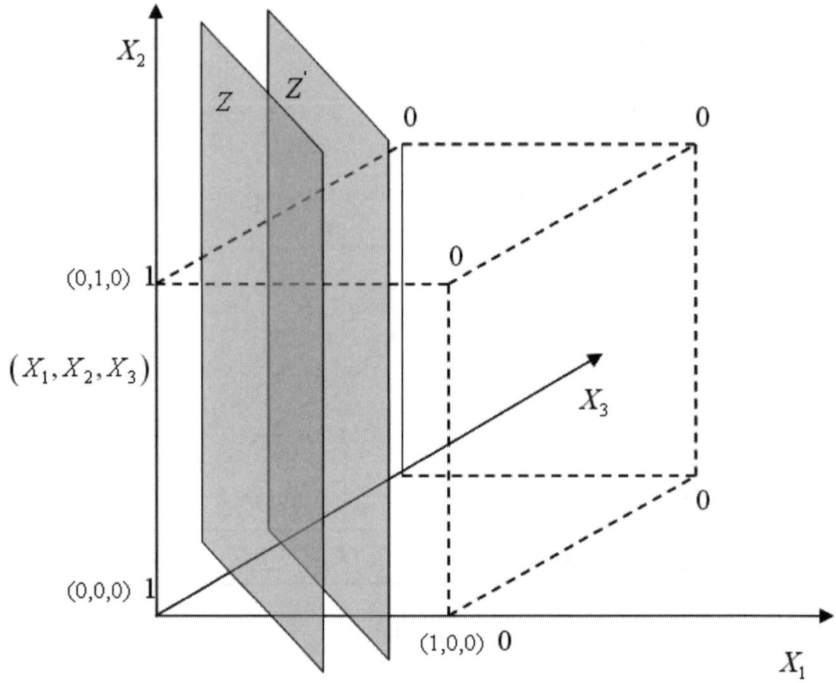

Figure 1.6.

In this example, we need to find a plane "Z" to separate the 6 "0" from the 2 "1". Similarly, it is clear that there are many hyperplanes "Z" can do the separation job. The complexity of the problem can be visualized when higher dimension input vector are considered. A training rule is needed to find the possible hyperplane to separate "1" from the "0".

In the above examples, whether it is a 3-dimensional cube or a 2-dimensional plane, they are all linearly separable cases. This means that we can use a linear line or a hyperplane to separate the "0" from the "1". In real world, problems may usually be linear non-separable. We use the famous "ex-OR" as example to illustrate the concept of linear non-separable. Fig. 1.7 shows that it is impossible to use a straight line to separate the "0" from the "1". This brings out the main shortcoming of a simple Perceptron model and the need of using multilayer Perceptron network which will all be briefed in later sections of this chapter.

Table 1.3

X_1	X_2	Output
0	0	0
0	1	1
1	0	1
1	1	0

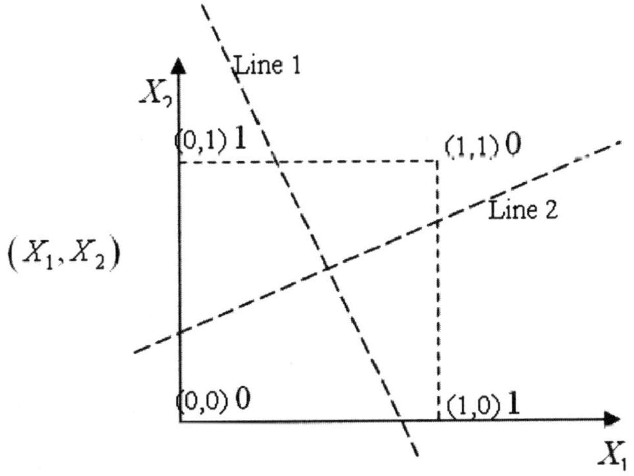

Figure 1.7.

In this case none of line "1", line "2" and all other linear lines can possible separate the 2 "0" from the 2 "1".

1.4.1.2. The Perceptron Model

The Perceptron model is the simplest type of neural network developed by Frank Rosenblatt in 1962. This type of simple network is rarely used now but it is significant in terms of its historical contribution to neural networks. A very simple form of Perceptron model is shown in Fig. 1.8. It is very much similar to the MCP model discussed in the last section. It has more than 1 inputs connected to the node summing the linear combination of the inputs connected to the node. Also, the resulting sum goes through a hard limiter which produces an output of +1 if the input of the hard limiter is positive. Similarly, it produces an output of -1 if the input is negative. It was first developed to classify a set of externally inputs into 2 classes of C_1 or C_2 with an output +1 signifies C_1 or C_2. Despite its early success, the single layer Perceptron network was proved by Minsky and Papert's work that it is unable to classify linear non-separable problem.

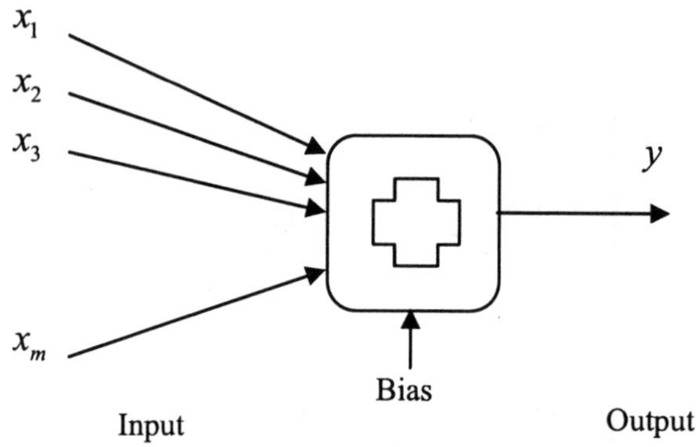

Figure 1.8. Single-layer Perceptron model

A 2-layer Perceptron network can be used for classifying linear non-separable problems. First, we consider the classification regions in a 2 dimensional space. Fig. 1.9 shows the 2-D input space in which a line is used to separate the 2 classes C_1 and C_2.

Similar to Eq. (1.2)

$$W_1X_1 + W_2X_2 - T = 0$$

The region below or on the right of the line is

$$W_1X_1 + W_2X_2 - T > 0 \tag{1.3}$$

Thus the region above or on the left of the line is

$$W_1X_1 + W_2X_2 - T < 0 \tag{1.4}$$

Fig. 1.10 shows that a linear non-separable case can be separated using a 2-layer perceptron model. Fig. 1.10(b) shows a 2-layer perceptron network separating linear non-separable case. This concept has paved the way for the later development of multi-layer feedforward model.

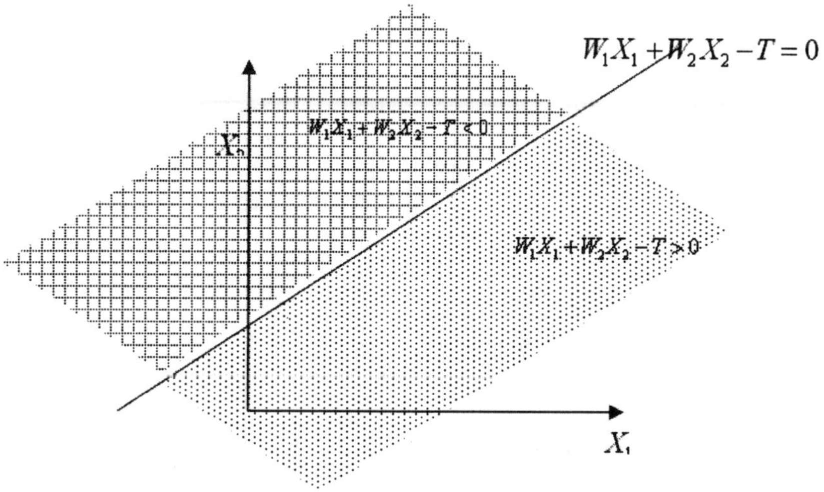

Figure 1.9.

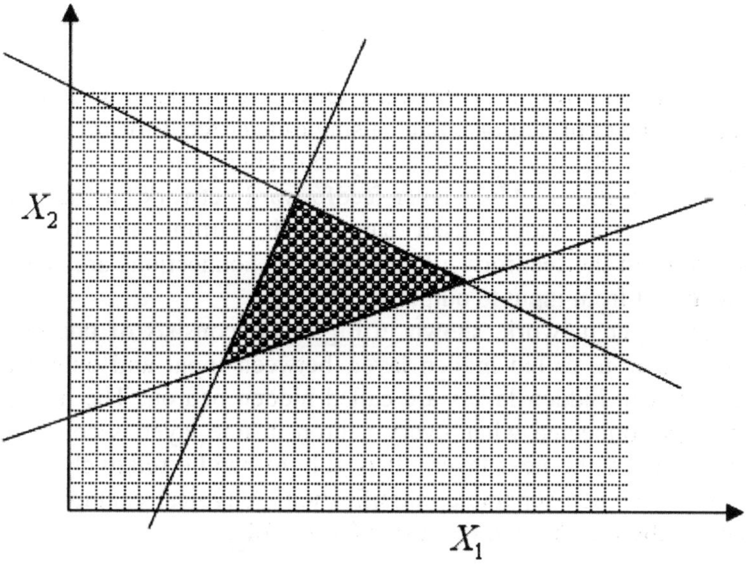

Figure 1.10. (a) It shows we can use 3-liear lines to separate linear non-separable case

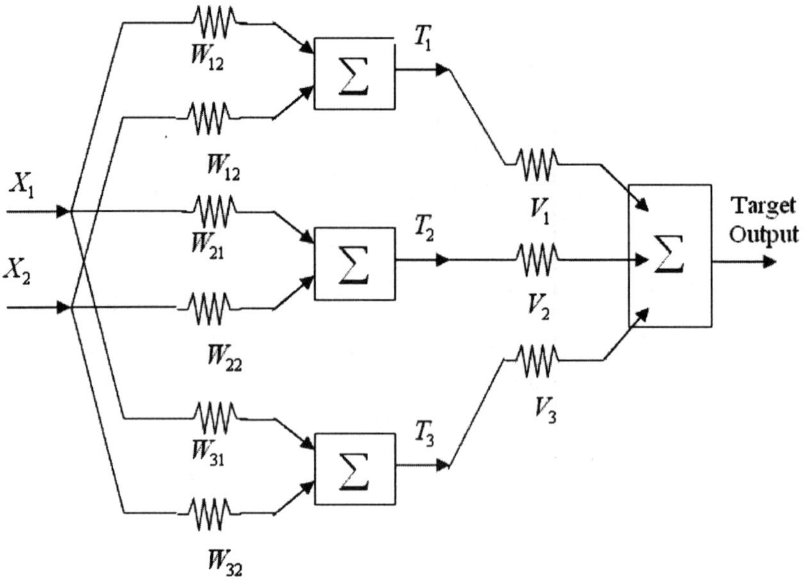

Figure 1.10. (b) A 2-layer perceptron network separating linear non-separable case

1.4.1.3. Multi-layer Feedforward Network

In a simple form, a feedforward network consists of an input layer and a single layer of neurons. Such a single-layer feedforward network is not capable of classifying nonlinearly separable patters as discussed in the previous sections. Multi-layer feedforward network has become the major and most widely used supervised learning neural network architecture. In the feedforward networks, all connections are acyclic or unidirectional from input to output layer.

Multi-layer network, shown in Fig. 1.11, consists of one or more layers of neuron, called hidden layer, between input and output layer. The neurons in hidden layers are called hidden neuron. The network is called fully connected when every neuron in one layer is connected to every neuron of the next layer. It is able to handle relatively complex task and linear non-separable classification. A typical example of such problem is the well known Exclusive OR (XOR).

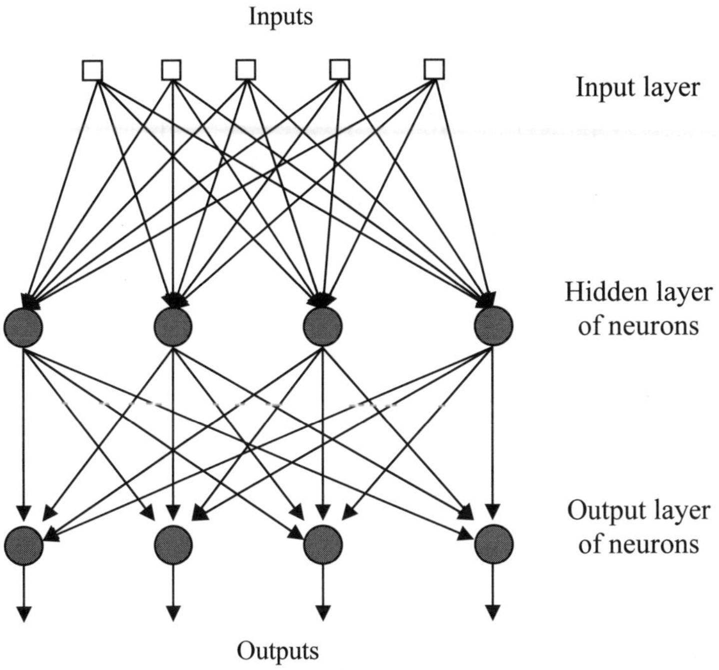

Figure 1.11. Multilayer feedforward network with 1 hidden layer

1.4.1.4. Recurrent Networks

Recurrent network is a special form of network. It can be a single layer or multiple hidden layers network. The basic difference from feedforward networks is that they have one or more feedback loops as shown in Fig. 1.12. The feedback loop can appear in many forms between any two neurons or layers. It typically involves unit delay element denoted by z^{-1}. Recurrent networks exhibit complex dynamics because they consist of a large number of feedforward and feedback connections. This characteristic provides them extra advantages in handling time-series related and dynamical problems over feedforward networks. Recurrent networks are also useful for processing special types of data such as graph-structured data. A recurrent network with a smaller network size may be equivalent to a complicated type of feedforward network architecture.

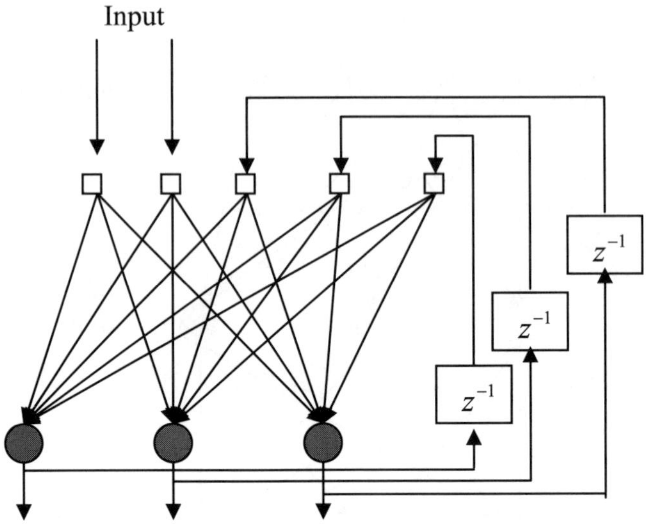

Figure 1.12. Recurrent network

There have been a wide applications of recurrent networks on intelligent control, system identification, and dynamical system applications. In these applications, theoretical study of stability, convergence of the network trajectory to the equilibrium, and their functional approximation ability are regarded important. These studies have widely been reported in many research articles. Some of these work include the proof of a discrete-time trajectory on a closed finite interval can be represented by using a discrete-time recurrent neural network. In the case of continuous-time recurrent networks, there is work showing that a continuous-time dynamical system without input i.e. $\dot{\mathbf{x}} = \mathbf{F}(\mathbf{x})$ can be approximated by a class of recurrent networks to an arbitrary degree of accuracy. Subsequently, there are work proving that a general dynamical continuous-time systems with control input, i.e. $\dot{\mathbf{x}} = \mathbf{F}(\mathbf{x},\mathbf{u})$ can be represented by recurrent networks. Different type of dynamical system like $\dot{\mathbf{x}} = \mathbf{f}(\mathbf{x}) + \mathbf{g}(\mathbf{x})\cdot\mathbf{u}$ has also been studied. All these works focus on the approximation problem of continuous-time recurrent networks to dynamical time-invariant system $\dot{\mathbf{x}} = \mathbf{F}(\mathbf{x},\mathbf{u})$. There are also study on dynamical time-variant systems, i.e. $\dot{\mathbf{x}}(t) = \mathbf{F}(\mathbf{x}, \mathbf{u}, t)$, which has wider industrial applications because most dynamical systems may be time-variant. Now the theoretical study of recurrent neural networks on approximating different dynamical systems has reached a very sophisticated stage that recurrent networks have been widely applied to many forecasting, functional approximation, system identification, and control problems. In recurrent network learning algorithm, special types of algorithms were designed to handle control problems. More approaches have been developed using recurrent neural networks based control methods. It has been pointed out that the fundamental shortcomings of current adaptive control techniques, such as nonlinear control laws which are difficult to derive, geometrically increasing complexity with the number of unknown parameters, and general unsuitability for real-time applications, have compelled researchers to look for solutions elsewhere. Recently, it has been shown that recurrent neural networks have emerged as a successful tool in the field of dynamical control systems. Funahashi and Nakamura have proved that any finite-time trajectory of a given n-dimensional dynamical system can be approximately realized by internal states of the output units of a continuous-time recurrent neural network when appropriate network topologies together with appropriate initial conditions are used. When recurrent networks are used to approximate and control an unknown nonlinear system through an on-line learning process, they may be considered as subsystems of an adaptive control system. The weights of the networks need to be updated using a dynamical learning algorithm during the control process. In establishing a convergence control action, many different recurrent training learning algorithms were developed, for instance Chow and Fang derived a 2-D nonlinear system mathematical model to describe the dynamic of a recurrent network and the dynamics of the learning process. The error dynamic equation is expressed in the form of a 2-D Roesser's model with time-varying coefficients. Based on 2-D system theory, a real-time iterative learning algorithm for trajectory tracking was derived.

1.4.2. Unsupervised Neural Networks

Unlike the supervised networks, unsupervised networks do not have a teacher in the training data set. The learning process of unsupervised neural networks is carried out from a self-organizing behavior. In the course of training, no external factor is used to affect the weights adjustment of the network. The correct outputs are not available during the course of training. For instance, a typical unsupervised network consists of an input layer and a competitive layer. Neurons on the competitive layer compete with each other via a simple competitive learning rule to best represent a given input pattern. Through competitive learning, the network output automatically reflects some statistical characteristics of input data such as data cluster, topological ordering etc.

Self-organizing map (SOM) is the most widely used unsupervised neural networks. Fig. 1.13(b) shows the architecture of a typical SOM network, in which all neurons, arranged on a fixed grid of the output layer, contain a weight vector similar to the input dimension. After the training, each neuron becomes representative of different types of input data. One of the most important characteristic of SOM lies in is its topological ordering which means that the neurons that have similar weight (in the input dimension) also close to each other in the SOM output map. This type of SOM map is useful in a many applications including clustering, visualization, quantization and retrieval.

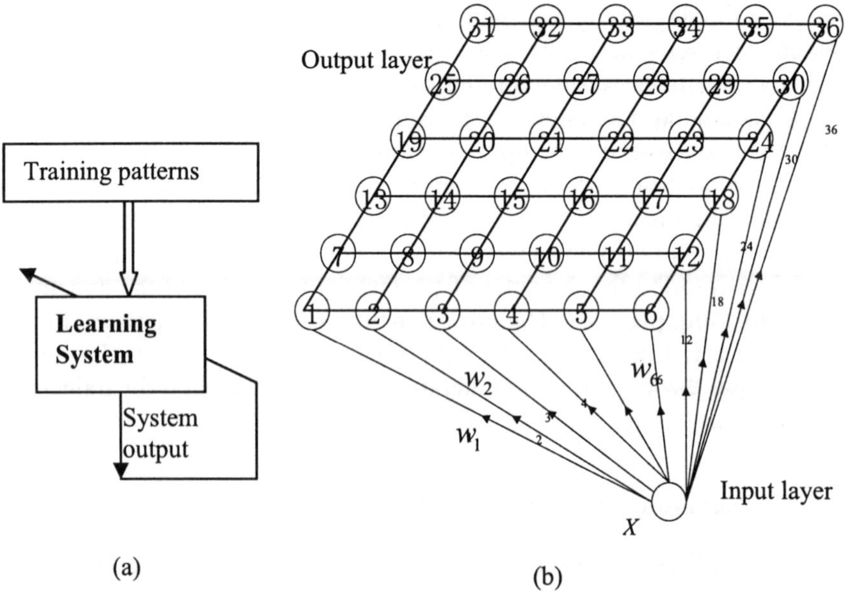

Figure 1.13. (a) Unsupervised learning process (b) SOM network architecture

1.5. Modeling and Learning Mechanism

One of the main functions of neural networks is about their excellent ability to model a complex multi-input multi-output system. Neural networks have widely been considered and used as a kind of soft mathematical modeling method. In a given high dimensional input output dataset, neural networks are able to provide a promising modeling service. Despite the fact that the neural learning procedures may be perceived rather straightforward from the perspective of application, every problem is vastly different because the problems usually consist of new measurements or test data that are not exactly the same as the training data seen amid the training period. For instance, consider a face detection problem in which the training dataset is given with a set of pictures containing faces. A picture can be represented by a vector of its pixel values in an n-dimensional vector. The training set may consist of an m pairs of data in which the label, which is the name of the face, associated with the picture is included. The training process should come up with a function h(x) which would deliver the correct label when given a query image. Usually, there are 2 main problems: (i) the function, h(x), may produce excellent results on the training set, but may perform rather poorly on new unseen pictures. (ii) New unseen pictures may never be identical as the previously seen images in the training set. Apparently, the task of training is not simply to reduce the training error. We are more concern on the generalization ability of the network.

There are many training algorithms developed in the past 10 years, they all share the similar goal of minimizing the training error in the shortest possible time. Despite their employed approaches may ranged from using gradient descent optimization to Least Squared mechanism, we need to analyze the problem from the perspectives of Information theory, and Statistics in order to come up with the best network for a given problem. In order to understand the fundamental of learning mechanism, we will start to illustrate it from a simple linear regression perspective.

1.5.1. Determination of Parameters

We can consider a simple linear regression example in which a linear line is used to model a given dataset as shown in Fig. 1.14. It investigates the electrical property of a new conducting material. The dots show how electrical current varies with the supply voltage. For a given supply voltage of 50 volts, we are asked to predict the magnitude of current flowing through the new material. This question can simply be answered by using a linear model. The simplest model is in the form

$$y = wx + b \tag{1.5}$$

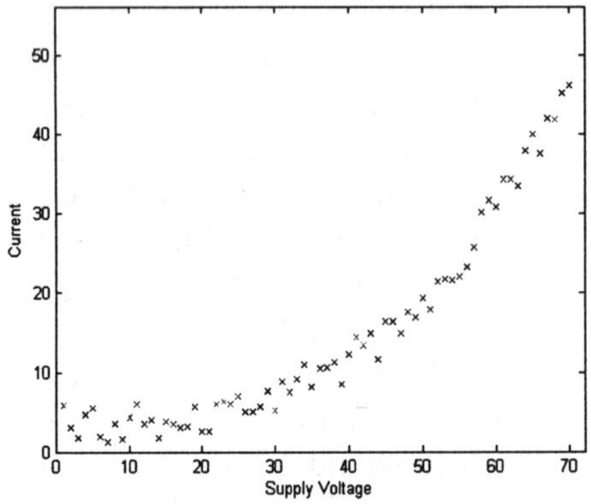

Figure 1.14.

The question becomes the determination of the 2 parameters, w and b. It is a question of finding the best straight line, which will be used as a predictor, drawn through the experimental data. In this simple linear regression question, the two parameters can be solved conventionally by using the Least Squared method.

$$w = \frac{\sum_i x_i d_i - (\sum_i x_i \sum_i d_i)/N}{\sum_i x_i^2 - (\sum_i x_i)^2 / N} \tag{1.6a}$$

$$b = \frac{\sum_i x_i^2 \sum_i d_i - \sum_i X_i \sum_i x_i d_i}{N \sum x_i^2 - (\sum_i x_i)^2} \tag{1.6b}$$

where d_i, x_i correspond to the measured current response, and voltage supply respectively, and N is the total number of data. In solving the linear regression problem, the 2 parameters are obtained by using the above equation. One can imagine that complexity of the problem when more parameters are involved. In neural network training mechanism, we have to rely on a more sophisticated optimization approach on solving the numerous parameters.

We use the same linear regression problem to illustrate the concept of a cost function approach. To make a "good predictor", we define a cost function E over the model parameters. One of the most widely used cost function for E is the sum squared error given

$$E = \frac{1}{2}\sum_i (d_i - y_i)^2 \tag{1.7}$$

The above equation is quadratic because of the square term. The performance surface is in a form of a convex surface, in which the minimum locates at the bottom. Fig. 1.15 shows how the sum squared error varies with the 2 parameters.

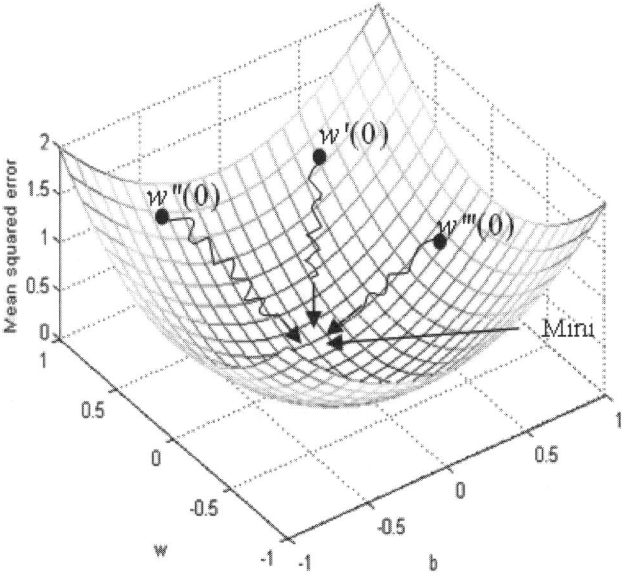

Figure 1.15.

The cost function E provides us an objective measure of predictive error for a specific choice of parameters. The best possible parameters can be obtained by minimizing the cost function or finding the minimum of the performance surface. We use gradient descent type optimization algorithm to find the minimum. This can easily be found as long as we are able to evaluate the gradient of the performance surface and follow its down hill path all the way along the surface. Different initial settings do not affect the final solution except providing different downhill paths. We can simply imagine that we are somewhere on or near the summit of a mountain. The shortest way to go down hill will simply to follow the steepest gradient of the terrain. But the selection of our step size is important when we go downhill because too large a step size will apparently be dangerous. In mathematical sense, too large a step size will likely cause oscillation around the minimum although it may speed up the optimization at the beginning. All these issues will be discussed in detail in later chapter of this book. The gradient descent optimization procedures are

1. Initialize the parameters randomly.
2. Evaluate the gradient of the error function with respect to each model parameter.
3. Adjust the model parameters by a certain step size in the direction of the steepest gradient.
4. Repeat steps 2 and 3 until the minimum is found.

Interestingly, the linear model shown in Eq. (1.5) can be implemented by the simplest form of a neural network shown Fig. 1.16. This simple network consists of a bias neuron, an input neuron, and a linear output neuron. In most feedforward neural network implementation, the bias neuron is set to have a constant input 1.

$$y_2 = x_1 W_{21} + (1)(W_{20}) \tag{1.8}$$

The weights W_{21} and W_{20} are determined using the gradient descent algorithm. When more weights are involved, the problem becomes a high dimensional one which cannot be depicted using a 3 dimensional space. But the overall mechanism can be perceived identical to the above described. It is worth noting that although the bias neurons and their weights are generally omitted from most architecture layout diagrams, all neurons consist of a bias weight with a constant input 1. The learning algorithm will determine the weight of the bias together with other weights.

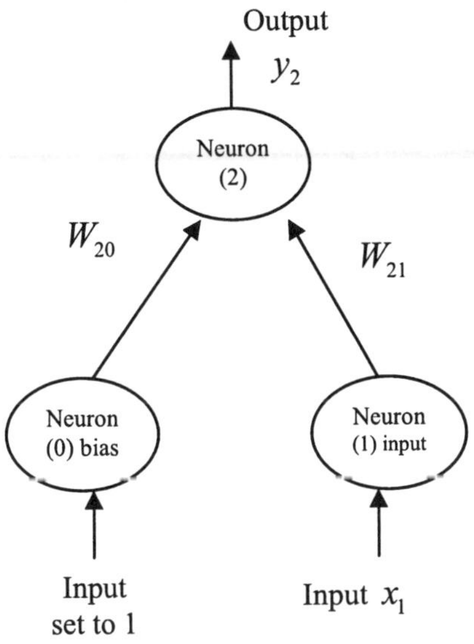

Figure 1.16.

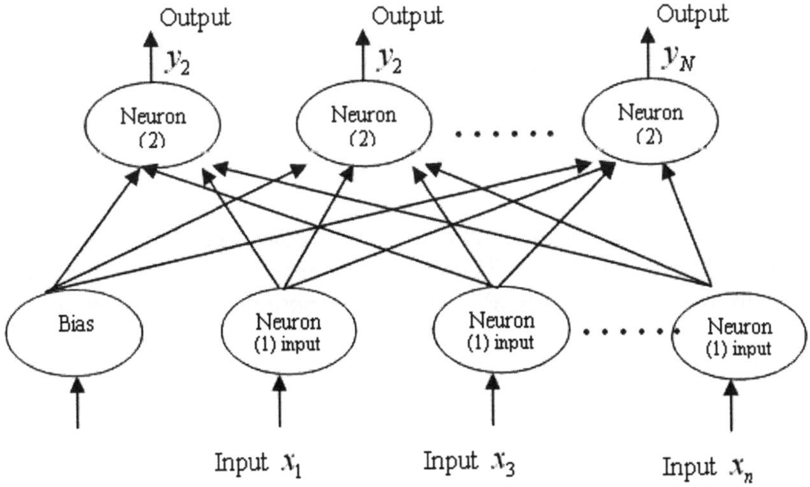

Figure 1.17.

The Current Vs Supply voltage example illustrate the determination of the current from any given supply voltage. This is a simple single input single output problem. If the problem is a more complicated consisting more than a single input and a single output variables, i.e. frequency, noise, bias, and phase, the above simple neural network model has to be modified by including more input and output neurons as shown in Fig. 1.17.

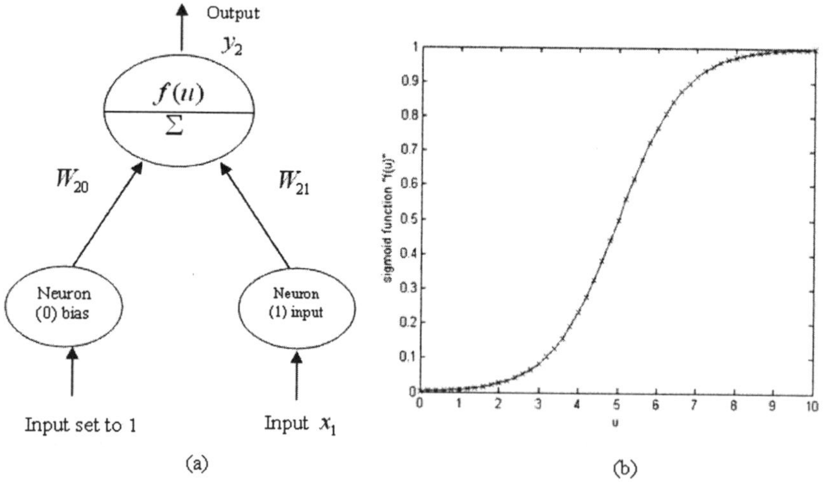

Figure 1.18. (a) A network with a neuron consisting of a nonlinear activation function $f(u)$ and (b) "sigmoid"- a typical activation function

We consider the curve fitting problem again from different perspective. In Fig. 1.14, it is clear that a curve will provide a much better fit than a linear line. Thus, it is reasonable to perform the curve fitting by including a nonlinear activation function inserted at the output of a neuron. An S-shaped sigmoid function $f(u)=\dfrac{1-e^{-\alpha u}}{1+e^{-\alpha u}}$ shown in Fig. 1.18(b) is one of the most widely used activation functions. The output of the network with a nonlinear activation function is

$$y=f(w_{21}x_1+w_{20}) \tag{1.9}$$

where $f(u) = \dfrac{1-e^{-\alpha u}}{1+\ e^{-\alpha u}}$. Here, u is the weighted input for the output neuron: $u = \left(w_{21}x_1 \ + \ w_{20}\right)$. This is the widely used feedforward neural network with nonlinear activation function.

1.5.2. Gradient Descent Searching Method

In the last section, we have shown why the performance surface is paraboloid when only 2 parameters are involved. In neural networks learning, a large number of weights are involved that results in a very complex high dimensional performance surface. Searching the performance surface using gradient descent method is an efficient way to determine the networks weights iteratively. As the search is to reach the minimum of the surface, the search direction must be opposite to the local gradient. Thus we can randomly initialize the weights as $W(0)$ where the index in the parentheses denotes the number of iteration, and W denotes the weight vector. With the given initial weights, we can evaluate the gradient of the surface at $W(0)$. The weights are thus adjusted by a magnitude proportional to the gradient and a step size. The new updated weights, $W(1)$ is then obtained. This process continues in a way

$$W(n+1)=W(n)-\eta\nabla J(n) \tag{1.10}$$

where η is a constant step size, and $\nabla J(n)$ denotes the gradient of the surface at nth iteration. The search mechanism stops when the stopping criteria are met. To improve the learning process without leading to oscillation, Rumelhant suggested the learning algorithm to be modified as

$$W(n+1)=W(n)-\eta\nabla J(n)+\alpha\Delta W(n-1) \tag{1.11}$$

where the α is a constant which determines the effect of the past weight changes on the current direction of movement in weight space. The constant α, called momentum, provides the damping effect and reduces the amount of oscillation during the course of training.

It is noted that most learning algorithms use a constant η during the whole course of learning. It is, however, obvious that a flexible learning rate that varies with different iterations can be a better learning mechanism. A relatively large learning rate at the beginning learning phase is employed but a relatively small learning rate has to be used in order to maintain learning stability when the search is close to the minimum. The ways of adjusting the learning rate according to different scenarios have never been straightforward. These involve complicated computation on the correlation between the weights change and their current learning rate. This will be discussed in later chapter of this book.

When we are handling a complex neural network, the performance surface is very complicated that cannot be depicted in a 3 dimensional space. We cannot be sure if there is only one minimum as shown in Fig. 1.15. Usually, the high dimensional performance surface may be very rugged that results in many local minima. One will never be sure that whether or not our searching is stuck in a local minimum. Also, we will never be able to obtain a zero error despite many trials. The searching mechanism is stopped when the magnitude of the error reaches a predefined level or when certain number of iteration is reached. This is called the stopping criteria. Trapping in local minima has also been a main concern to many neural network users. The likelihood of getting stuck in local minima increases when the complexity of a problem increases because the high dimensional performance surface may become very rugged that results in many local minima. One of the ways to relieve this problem is using advanced learning algorithms or even employing different cost function that may have an effect of changing the performance surface. These issues will be detailed in later chapters of this book.

Exercises

Q1.1. A two input binary neuron shown in Fig. 1.19 has a unit step activation function with bias = 0.5. Find the space of possible values of weights of α, and β for input x_1 and x_2 respectively if the neuron is

1) OFF for input (1.0, 1.0);
2) OFF for input (0.5, -1.0);
3) ON for input (0.5, -0.5).

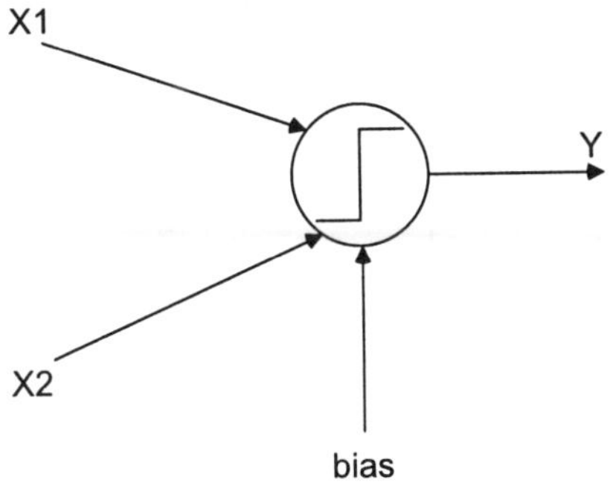

Figure 1.19.

Q1.2. The gradient vector of a function $g(x_1, x_2) = 12x_1^2 + 3x_2^2$ is defined by a column vector $\nabla g = [\frac{\partial g}{\partial x_1} \frac{\partial g}{\partial x_2}]^T = c$. It is given that the normalized gradient vector $\overline{c} = \frac{c}{\sqrt{c^T c}}$. Use gradient descent method to search the minimum for $g(x_1, x_2)$ with a given initialization of $x(0) = [1 3]^T$. Show and find the updated values $x(1)$ and $x(2)$, with a step size of 0.5.

Q1.3. An LMS algorithm is used to implement a dual-input, single-weight adaptive noise canceller as shown in Fig. 1.20. Use the LMS rule to adjust the weight w of the adaptive filter. Set up an updating equation relating $w(n+1)$ to $w(n), y(n), x(n)$ and the learning rate η.

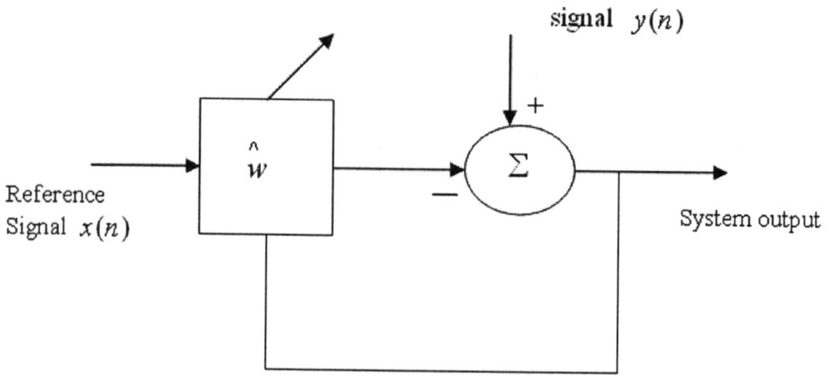

Figure 1.20.

Q1.4. Write a simple program to generate a dataset relating y to x1, x2, x3, and x4 as the following. Use a feedforward network to model the following equation. Use 100 data for training and another 50 unseen data as testing.

$$y = 5 + 4x_1 + 3X_2 - 2x_1x_4 + 2x_3^{\ 2}$$

Q1.5. Fig. 1.21 and Fig. 1.22 show the simple McCulloch and Pitts (MCP) model used for modeling a logic function "AND". It shows the separation line with a gradient of - 0.7, which intersects the y-axis at 1.3, where x_1 is denoted by the y axis, and x_2 is denoted by the x axis. Assuming the threshold T is 1, determine the weights w_1 and w_2 for input x_1 and x_2 respectively.

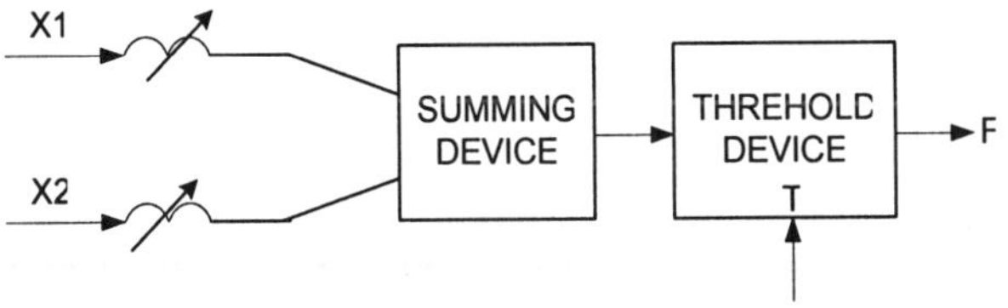

Figure 1.21.

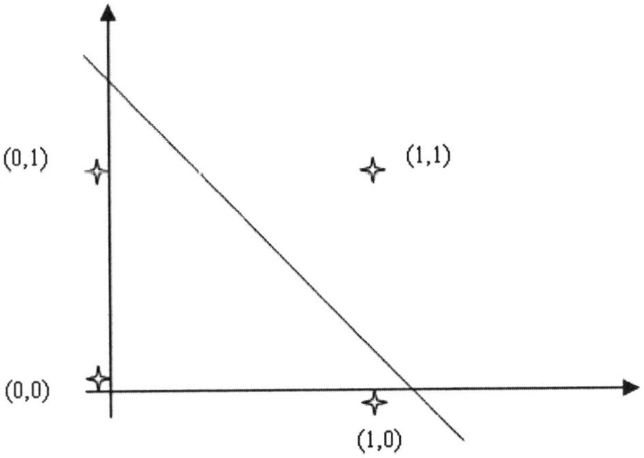

Figure 1.22.

Chapter 2

Learning Performance and Enhancement

Neural networks structures, broadly classified as feedforward (without feedback) and recurrent (involving feedback), have numerous processing elements (activation functions) and connections (weights). Most network topologies are in the forms of layer that categorize single-layer networks and multi-layer networks. The single-layer network is the simplest form that an input layer of source nodes projects onto an output layer of neurons. The multi-layer network is mostly useful because the presence of one or more hidden layers (connected between input layer and output layer) enable the extraction of higher-order statistics for the neural network to acquire a global perspective. It is particularly valuable when the problems are rather difficult and diverse. Interconnections exist between each layer of the layered type neural networks. The strengths of the interconnections are known as synaptic weights that are used to store the knowledge acquired by a neural network through a learning process. The procedure used to perform such a learning process is called a learning algorithm.

Learning algorithm can be defined as a function to adjust the synaptic weights of a neural network in an orderly fashion so as to attain a desired objective. Since the past few decades, different types of learning algorithms have been developed by many researchers. For instances, Hebbian learning and competitive learning were developed for unsupervised learning, whereas, least-mean-squares (LMS) algorithm and error Backpropagation (BP) algorithm (Rumelhart et al., 1986) were developed for supervised learning. All these classic works have laid an immense contribution to the modern neural network research. They are much easier to analyze theoretically and can often be implemented efficiently in a variety of applications, such as time-series forecasting, pattern classification, system identification, and robotics and visions applications. Since the flourish of these early works, the supervised learning method, especially for the error BP, has been widely used for multilayer neural networks.

This chapter is organized as follows: Section 2.1 presents the fundamental of gradient descent optimization. Section 2.2 presents the Back-propagation learning algorithm. The derivation are briefly described, however, the limitations are also addressed in this section. Section 2.3 focuses on the convergence performance in terms of learning speed. Three algorithms are described to increase the rate of convergence. Following that, section 2.4 presents the algorithm initialization has a significant affect to the learning performance. The methods of initialization enhancement are described and suggested how the networks weights can be set effectively in this section. The problem of local minima is another issue to affect the convergence which is addressed in section 2.5. This section describes several global optimization methods which can be used to train neural networks effectively. Some evaluations are also presented in this section. Finally, section 2.6 presents the concluding remarks of this chapter.

2.1. Fundamental of Gradient Descent Optimization

In order to train a neural network by gradient descent, we need to be able to compute the gradient ∇E of the cost function with respect to each weight w_{ji} of the network. It tells us how a small change in that weight will affect the overall error E. We begin by splitting the cost function into separate terms for each pattern p in the training data:

$$E = \sum_p E^p, \qquad E^p = \frac{1}{2}\sum_j \left(t_j^p - y_j^p\right)^2 \tag{2.1}$$

where j ranges over the output units of the network. Since differentiation and summation are interchangeable, we can split the gradient into separate components for each training pattern:

$$\begin{aligned}\nabla E &= \frac{\partial E}{\partial w_{ji}} \\ &= \frac{\partial}{\partial w_{ji}} \sum_p E^p \\ &= \sum_p \frac{\partial E^p}{\partial w_{ji}}\end{aligned} \tag{2.2}$$

In what follows, we describe the computation of the gradient for a single data point, omitting the superscript p in order to make the notation easier to follow.
First use the ***chain rule*** to decompose the gradient into two factors:

$$\frac{\partial E}{\partial w_{ji}} = \frac{\partial E}{\partial y_j} \frac{\partial y_j}{\partial w_{ji}} \tag{2.3}$$

The first factor can be obtained by differentiating eq. (2.1) above:

$$\frac{\partial E}{\partial y_j} = -\left(t_j - y_j\right) \tag{2.4}$$

Using $y_j = \sum_i w_{ji} y_i$, the second factor becomes

$$\begin{aligned}\frac{\partial y_j}{\partial w_{ji}} &= \frac{\partial}{\partial w_{ji}} \sum_i w_{ji} y_i \\ &= y_i\end{aligned} \tag{2.5}$$

Putting the equations (2.3)-(2.5) back together, we obtain

$$\frac{\partial E}{\partial w_{ji}} = -\left(t_j - y_j\right) y_i \tag{2.6}$$

To find the gradient ∇E for the entire data set, we sum at each weight the contribution given by equation (2.6) over all the data points. We can then subtract a small proportion μ (called the learning rate) of ∇E from the weights to perform gradient descent. The gradient descent algorithm is sum up as follows:

1. Initialize all weights to small random values
2. REPEAT until done

a. For each weight w_{ji} set $\Delta w_{ji} := 0$

b. For each data pattern $(\mathbf{x},\mathbf{t})^P$

   i. Set input units to **x**

   ii. Compute value of output units

   iii. For each weight w_{ji} set $\Delta w_{ji} := \Delta w_{ji} + (t_j - y_j)y_i$

c. For each weight w_{ji} set $w_{ji} := w_{ji} + \mu \Delta w_{ji}$

The algorithm terminates once we are at, or sufficiently near to, the minimum of the error function, where $\nabla E = 0$. We say then that the algorithm has ***converged***.

In this gradient descent algorithm, an important consideration is the learning rate μ, which determines by how much we change the weights *w* at each step. If μ is too small, the algorithm will take a long time to converge (see Fig. 2.1a). Conversely, if μ is too large, we may end up bouncing around the error surface out of control – the algorithm diverges (see Fig. 2.1b). This usually ends with an overflow error in the computer's floating point arithmetic.

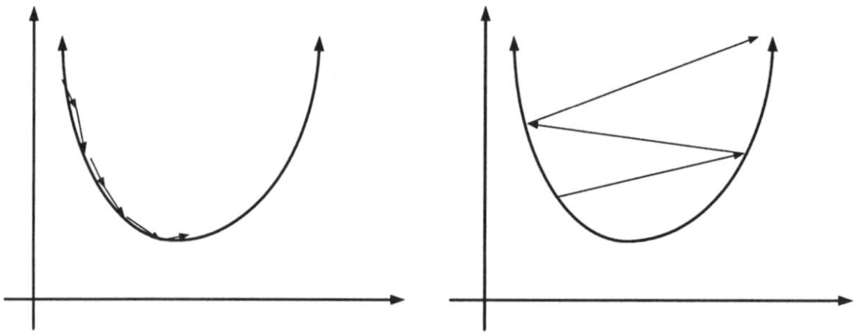

(a) (b)

Figure 2.1. Learning rate for gradient descent optimization, (a) Small learning rate – Slow convergence, (b) Large learning rate – Divergence

As described in the above we have accumulated the gradient contributions for all data points in the training set before updating the weights. This method is often referred to as batch learning. An alternative approach is online learning, from which weights are updated immediately after seeing each data point. Since the gradient for a single data point can be considered a noisy approximation to the overall gradient ∇E (Fig. 2.2), this is also called ***stochastic*** (noisy) gradient descent.

Online learning has a number of advantages:

- it is often much faster, especially when the training set is ***redundant*** (contains many similar data points),
- it can be used when there is no fixed training set (new data keeps coming in),
- it is better at tracking ***nonstationary*** environments (where the best model gradually changes over time),
- the noise in the gradient can help to escape from local minima (which are a problem for gradient descent in nonlinear models).

These advantages are, however, bought at a price: many powerful optimization techniques, such as, conjugate and second-order gradient methods, support vector machines, Bayesian methods, etc. – are batch methods that cannot be used online. A compromise between batch and online learning is the use of "mini-batches": the weights are updated after every *n* data points, where *n* is greater than 1 but smaller than the training set size.

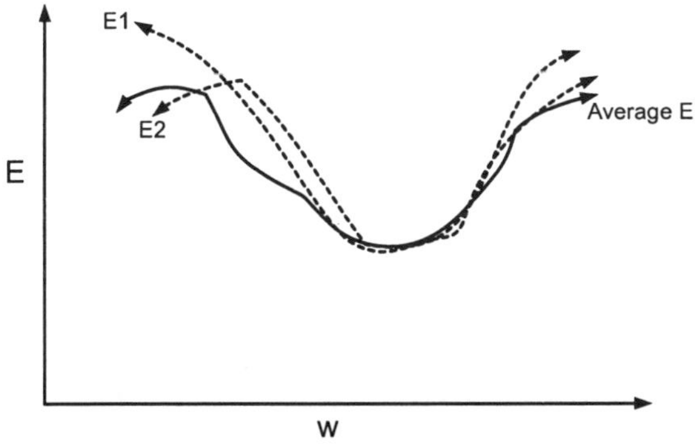

Figure 2.2. A noisy approximation to the overall gradient

2.2. Conventional Backpropagation Algorithm

Recently, multilayer neural networks have been applied successfully to solve lots of difficult and diverse problems through employing various supervised learning procedures among which the error Backpropagation (BP) learning algorithm appears to be the most popular. This algorithm is an iterative gradient based algorithm proposed to minimize an error between the actual output vector of the network and the desired output vector. Because of the nonlinearity of neural models, algorithms for supervised training are mainly based upon the nonlinear optimization methods. In the following, we will describe the batch-mode training using backpropagation algorithm.

For notation convenience, the layers are numbered from bottom up beginning with 1. Analogous to the single-layer perceptron, layer 1 consists of fanout processing neurons that simply accept the individual elements of the input patterns and pass them directly to all units of layer 2. Each neuron on other layers receives the weighted signal from each of the neurons of the layer below. After the summation and the activation function operations, the output is distributed to all neurons of the upper layer next to this layer. A multilayer feedforward neural network, showing the notation for neurons and weights is shown in Fig. 2.3.

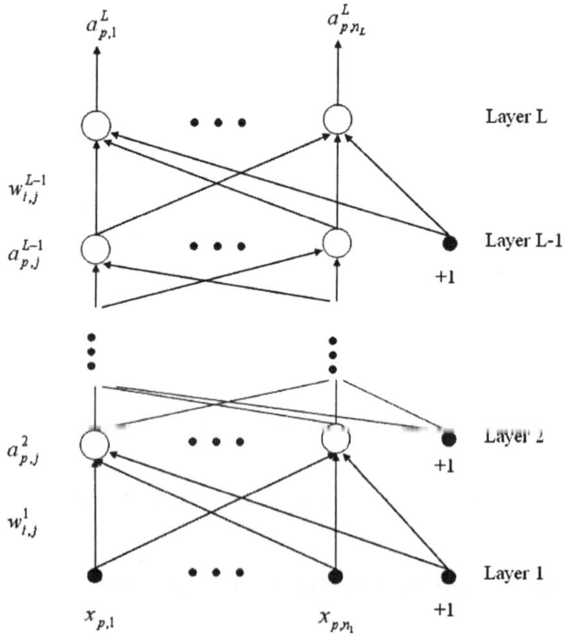

Figure 2.3. A multilayer feedforward neural network, showing the notation for neurons and weights

Let $x_{p,i}$ be the current input pattern for the *i*th neuron of the input layer, $a_{p,j}^{l}$ be the corresponding output of the *j*th neuron of layer *l*, and $w_{i,j}^{l}$ be the weight connecting the *i*th neuron of layer *l* and the *j*th neuron of layer $l+1$, the computation procedure can be described as

$$a_{p,n_1+1}^{1} = 1$$

and $$a_{p,i}^{1} = x_{p,i} \qquad i = 1,2,...,n_1 \tag{2.7a}$$

$$o_{p,j}^{l} = \sum_{i=1}^{n_l+1} a_{p,i}^{l} w_{i,j}^{l} \tag{2.7b}$$

$$a_{p,j}^{l+1} = f(o_{p,j}^{l})$$

$$p = 1,...,P, l = 1,2,...,L-1 \text{ and } j = 1,2,...,n_{l+1} \tag{2.7c}$$

$$a_{p,n_{l+1}+1}^{l+1} = 1 \;\; p = 1,...,P, l = 1,2,...,L-2 \tag{2.7d}$$

where n_l denotes the total number of neurons on layer *l*, *L* denotes the number of layers in the architecture, and P denotes the number of training patterns. $a_{p,n_{l+1}+1}^{l+1}$ are set to 1 for making the weight $w_{n_{l+1}+1,j}^{l+1}$ as a threshold for the *j*th neuron on layer $l+1$. The outputs of the final layer are evaluated by propagating the signals in this way.

The operation of the network described above can equally well be represented by matrix equations. For a multilayer feedforward neural network with *L* layers, the layer *l* consists of $n_l + 1$ neurons ($l = 1,...,L-1$) in which the last neuron is a bias node with a constant output of 1. The first layer with $n_1 + 1$ neurons is the input layer, the *L*-th one with n_L neurons is the output layer. All the given inputs can be represented by a matrix A^1 with P rows and $n_1 + 1$ columns. All entries of the last column of the matrix A^1 are constant 1. Similarly, the outputs of the layer *l* can be represented by A^l with P rows and $n_l + 1$ columns. The outputs of the last layer can be represented by A^L with P rows and n_L columns. Similarly, the target can be represented by a matrix T^L

with P rows and n_L columns. The weights between neurons in the layers l and $l+1$ form a matrix W^l with entries $w_{i,j}^l$ $(i=1,...,n_l+1$, $j=1,...,n_{l+1})$. Entry $w_{i,j}^l$ connects neuron i of layer l with neuron j of layer $l+1$.

The output of all hidden layers and the output layer are obtained by propagating the training patterns through the network. Let us define the matrix

$$O^l = A^l W^l \tag{2.8a}$$

The entries of A^{l+1} for all layers (i.e., $l=1,...,L-1$) are evaluated as follows:

$$a_{p,j}^{l+1} = f(o_{p,j}^l)$$

$$p=1,...,P \text{ and } j=1,2,...,n_{l+1} \tag{2.8b}$$

In this way, the patterns propagate through the network until the outputs of final layer are computed. An algorithm is required to adjust the weights so that the network learns how to map the input patterns to the output patterns. The most widely used algorithm for training feedforward neural networks is the backpropagation algorithm.

Learning is achieved by adjusting the weights such that the network output, A^L is as close as possible or equal to the target, T^L. The most frequently used error function for measuring the difference between the output and the target is the mean squared error function. It is defined as

$$E = \frac{1}{2P}\sum_{p=1}^{P}\sum_{j=1}^{n_L}(t_{p,j} - a_{p.j}^L)^2 \tag{2.9}$$

Explicit solution of the weights is impossible to obtain because of the nonlinearity of the network. Iterative methods are required to solve this problem. Backpropagation algorithm is the classical algorithm for training a multilayer perceptron. This algorithm is based on a gradient-descent method to minimise the error cost function; i.e., the changes of weights are proportional to the error gradient. Mathematically,

$$\Delta w_{i,j}^l = -\eta \frac{\partial E}{\partial w_{i,j}^l} \tag{2.10}$$

where η is the learning rate. The weight changes $\Delta w_{i,j}^{L-1}$ for the weights connecting to the final layer are obtained by

$$\Delta w_{i,j}^{L-1} = -\frac{\eta}{2P}\sum_{p=1}^{P}\sum_{j=1}^{n_L}\frac{\partial}{\partial w_{i,j}^{L-1}}\left(t_{p,j} - a_{p,j}^{L}\right)^2 \tag{2.11}$$

Notice that for a given j, only $a_{p,j}^{L}$ has a relation with $w_{i,j}^{L-1}$, we get

$$\Delta w_{i,j}^{L-1} = \frac{\eta}{P}\sum_{p=1}^{P}\left(t_{p,j} - a_{p,j}^{L}\right)\frac{\partial a_{p,j}^{L}}{\partial w_{i,j}^{L-1}} \tag{2.12}$$

The partial derivative $\partial a_{p,j}^{L} / \partial w_{i,j}^{L-1}$ can be evaluated using the chain rule

$$\frac{\partial a_{p,j}^{L}}{\partial w_{i,j}^{L-1}} = \frac{\partial a_{p,j}^{L}}{\partial o_{p,j}^{L-1}}\frac{\partial o_{p,j}^{L-1}}{\partial w_{i,j}^{L-1}} \tag{2.13}$$

The two factors are obtained as follows:

$$\frac{\partial a_{p,j}^{L}}{\partial o_{p,j}^{L-1}} = f'\left(o_{p,j}^{L-1}\right) \tag{2.14}$$

and

$$\frac{\partial o_{p,j}^{L-1}}{\partial w_{i,j}^{L-1}} = \frac{\partial}{\partial w_{i,j}^{L-1}}\sum_{i=1}^{n_{l-1}+1} a_{p,i}^{L-1} w_{i,j}^{L-1} = a_{p,i}^{L-1} \tag{2.15}$$

From Eqns. (2.11-2.12)

$$\begin{aligned}\Delta w_{i,j}^{L-1} &= \frac{\eta}{P}\sum_{p=1}^{P}\left(t_{p,j} - a_{p,j}^{L}\right)f'\left(o_{p,j}^{L-1}\right)a_{p,i}^{L-1} \\ &= \frac{\eta}{P}\sum_{p=1}^{P}\delta_{p,j}^{L-1} a_{p,i}^{L-1}\end{aligned} \tag{2.16}$$

where $\delta_{p,j}^{L-1}$ is defined as

$$\delta_{p,j}^{L-1} = \left(t_{p,j} - a_{p.j}^{L}\right) f'(o_{p,j}^{L-1}) \tag{2.17}$$

The weights change for the weights connected to the second to the last layer are obtained from the following equations:

$$\begin{aligned}
\Delta w_{i,j}^{L-2} &= -\eta \frac{\partial E}{\partial w_{i,j}^{L-2}} \\
&= -\eta \sum_{p=1}^{P} \frac{\partial E}{\partial o_{p,j}^{L-2}} \frac{\partial o_{p,j}^{L-2}}{\partial w_{i,j}^{L-2}} \\
&= -\eta \sum_{p=1}^{P} \frac{\partial E}{\partial o_{p,j}^{L-2}} a_{p,i}^{L-2} \\
&= -\eta \sum_{p=1}^{P} \frac{\partial E}{\partial a_{p,j}^{L-1}} \frac{\partial a_{p,j}^{L-1}}{\partial o_{p,j}^{L-2}} a_{p,i}^{L-2} \\
&= \eta \sum_{p=1}^{P} \left(- \frac{\partial E}{\partial a_{p,j}^{L-1}} \right) f'\left(o_{p,j}^{L-2}\right) a_{p,i}^{L-2}
\end{aligned} \tag{2.18}$$

By applying the chain rule, $\partial E / \partial a_{p,j}^{L-1}$ can be evaluated. Specifically, we write

$$\begin{aligned}
-\frac{\partial E}{\partial a_{p,j}^{L-1}} &= -\sum_{k=1}^{n_L} \frac{\partial E}{\partial o_{p,k}^{L-1}} \frac{\partial o_{p,k}^{L-1}}{\partial a_{p,j}^{L-1}} \\
&= \frac{1}{P} \sum_{k=1}^{n_L} \delta_{p,k}^{L-1} w_{j,k}^{L-1}
\end{aligned} \tag{2.19}$$

We finally get the weight change

$$\Delta w_{i,j}^{L-2} = \frac{\eta}{P} \sum_{p=1}^{P} \left(\sum_{k=1}^{n_L} \delta_{p,k}^{L-1} w_{j,k}^{L-1} \right) f'\left(o_{p,j}^{L-2}\right) a_{p,i}^{L-2} \tag{2.20}$$

If we denote

$$\delta_{p,j}^{L-2} = \left(\sum_{k=1}^{n_L} \delta_{p,k}^{L-1} w_{j,k}^{L-1} \right) f'\left(o_{p,j}^{L-2}\right) \tag{2.21}$$

The weight change $\Delta w_{i,j}^{L-2}$ can then be simplified as

$$\Delta w_{i,j}^{L-2} = \frac{\eta}{P} \sum_{p=1}^{P} \delta_{p,j}^{L-2} a_{p,i}^{L-2} \tag{2.22}$$

From the Eq. (2.21), it is noticed that the deltas at an internal node can be evaluated in terms of the deltas at an upper layer. Thus, starting at the highest layer – the output layer – we can evaluate $\delta_{p,j}^{L-1}$ using the Eq. (2.17), and we can then propagate the "error" backward to lower layers. The operation of the algorithm justifies the term "backpropagation".

By analogy, we can show that the weights change for other lower layers of weights,

$$\Delta w_{i,j}^{l} = \frac{\eta}{P} \sum_{p=1}^{P} \delta_{p,j}^{l} a_{p,i}^{l} \qquad l = 1, \cdots, L-1 \tag{2.23}$$

and

$$\delta_{p,j}^{l} = \left(\sum_{k=1}^{n_{l+2}} \delta_{p,k}^{l+1} w_{j,k}^{l+1} \right) f'\left(o_{p,j}^{l}\right) \qquad l = 1, \cdots, L-2 \tag{2.24}$$

The learning procedure therefore consists of the network starting with a random set of weight values, choosing one of the training patterns and evaluating the output(s) using that pattern as input in a feedforward manner. Using the backpropagation procedure, all the weight changes for that pattern are evaluated. This procedure is repeated for all the patterns in the training set so that the $\Delta w_{i,j}$ for all the weights are obtained. The corrections to the weights are made. One epoch is completed.

The choice of the value of learning rate is important when we implement a neural network. A large learning rate corresponds to rapid learning but might also result in oscillations. In 1986, Rumelhart, Hinton, and Williams (1986) suggested that the expression was modified to include a momentum term. That is,

$$\Delta w_{i,j}^{l}(n+1) = -\eta \frac{\partial E}{\partial w_{i,j}^{l}} + \alpha \Delta w_{i,j}^{l}(n) \tag{2.25}$$

where the subscript n indexes the presentation number, and α is a constant which determines the effect of past weight changes on the current direction of movement in the weight space. This provides a kind of momentum in weight space that effectively filters out high-frequency variations of the error-surface in the weight space. This is useful in spaces containing long ravines that are characterized by sharp curvature across the ravine and a gently sloping floor. The sharp curvature tends to cause divergent oscillations across the ravine. To prevent these it is necessary to take very small steps, but this causes very slow progress along the ravine. The momentum filters out the high curvature and thus allows the effective weight steps to be bigger. The values of η and α for rapid training are dependent on the training patterns and the architecture of the neural network. Although the backpropagation algorithm with momentum term can speed up the training process, it is too slow for practical applications. This is one of the difficulties for training neural networks by the backpropagation algorithm.

Another difficulty of the backpropagation algorithm in practice is that the learning procedure is always trapped in a local minimum. In 1992, Gori and Tesi (1992) proposed a theoretical framework of the Backpropagation algorithm to identify the problem of local minima. In their studies, they illustrated that the gradient descent type BP learning algorithm are prone to local minima. For any nonlinear optimization problem, it is hardly to guarantee that the optimal solution is obtained by such numerical solving schemes. The major reason is with the intrinsic shape of the error surfaces which is normally fixed and independent of the learning algorithm. Frasconi et al. deduced several theorems stating that the cost function has a unique minimum if the patterns of the learning environment are only linearly separable. From the theoretical point of view, their results corroborate that the BP algorithm is not a very reliable learning algorithm in most practical applications. Also, there are very useful examples illustrated by Gori and Tesi (1990) trying to show that the BP algorithm always easily gets stuck in local minima.

2.3. Convergence Enhancement

In order to reduce the time taken to train feedforward neural networks, Sutton (1986) proposed that each weight has its own learning rate. The learning rate of each weight should increase or decrease its value according to the number of sign changes observed in the partial derivative of the error function with respect to the corresponding weight. On the basis of the above heuristics, Silva and Almeida (1990) developed an efficient algorithm with locally adaptive learning rate. This algorithm will be compared with the extended backpropagation algorithm in the following section. This algorithm employs batch back-propagation with momentum. Consider that in a given learning process the sign of a certain component of the gradient remains equal for several iterations. This fact suggests that the error surface has a smooth variation along the axis and, therefore, the learning rate for this particular component should be increased. On the other hand, if the sign of some component changes in several consecutive iterations, the learning rate parameter should be decreased to avoid oscillation.

A simple and effective way of implementing this basic idea is as follows. The weight change for the synapse linking neurons i and j at epoch n as

$$\Delta w_{i,j}(n) = \eta_{i,j}(n) v_{i,j}(n) \tag{2.26}$$

where $v_{i,j}(n)$ is given by

$$v_{i,j}(n) = -\frac{\partial E}{\partial w_{i,j}} + \beta v_{i,j}(n-1) \tag{2.27}$$

where β is the momentum coefficient. The momentum term smoothes the gradient vector in an adaptive way and amplifies the gradient component along the valley direction. The learning rate parameters are adapted as follows:

$$n_{i,j}(n) = \begin{cases} u\eta_{i,j}(n-1) & \text{if} \quad \dfrac{\partial E(n)}{\partial w_{i,j}} \dfrac{\partial E(n-1)}{\partial w_{i,j}} > 0 \\ d\eta_{i,j}(n-1) & \text{if} \quad \dfrac{\partial E(n)}{\partial w_{i,j}} \dfrac{\partial E(n-1)}{\partial w_{i,j}} < 0 \end{cases} \tag{2.28}$$

Their recommended values of the parameters are:

$$1.1 \le u \le 1.3, \qquad 0.75 \le d \le 0.9 \quad \text{and} \quad u \approx d^{-1} \tag{2.29a}$$

$$\eta_{i,j}(0) = 10^{-3}, \qquad \beta = 0.1 \tag{2.29b}$$

The values of u and d are obtained after performing a wide range of tests. For the d parameters, a value slightly below $1/u$ enables the adaptive process give a small preference to learning rate decrease, yielding a somewhat more stable convergence process.

As the fast increase in learning rate may drive the learning process to an unstable state, the following procedure is used to avoid this. If the present error is greater than that in the previous iteration, the new weight changes are rejected. However, the adaptation of the learning rate is still performed as usual, the gradient evaluated at the rejected iteration point is used as a reference. With this technique, a valid iteration point can generally be obtained after a few training epochs. If this strategy does not work after a few trials, it is always possible to simply reduce all the learning rate parameters by a fixed factor and repeat the process.

Despite the fact that the adaptive backpropagation algorithm can reduce the number of iterations to converge, the overall time taken for the training of rather small networks with only a few thousand weights is still unacceptably long. In the following, three new training algorithms will be described. These algorithms are developed to further reduce the time taken for training.

2.3.1. Extended Backpropagation Algorithm

In the backpropagation algorithm, training is only based on the current downhill gradient of the error surface $-\nabla E(t)$ and last change in weight $\Delta W(t-1)$. The two column vectors, $-\nabla E(t)$ and $\Delta W(t-1)$, are n dimensional vectors with elements $-\frac{\partial E(t)}{\partial w_{i,j}^{l}}$ and $\Delta w_{i,j}^{l}(t-1)$ respectively at the t-th iteration, and n is the total number of weights in the neural network. The local topography of error surface can be thoroughly evaluated for determining the optimum learning rate and momentum coefficient. The higher derivatives of the error function and line search technique are often used to find the optimum learning rate; however, a long computational time and a larger storage is required for these processes. In the EBP algorithm, the learning rate adaptation is simply based on the correlation coefficient between the current downhill direction of the local gradient and the previous weight update. The correlation coefficient between the negative gradient and last weight update is given by

$$r(t)=\frac{\sum_i\sum_j\sum_l\left(-\nabla E_{i,j}^l(t)-\overline{-\nabla E_{i,j}^l(t)}\right)\left(\Delta w_{i,j}^l(t-1)-\overline{\Delta w_{i,j}^l(t-1)}\right)}{\left[\sum_i\sum_j\sum_l\left(-\nabla E_{i,j}^l(t)-\overline{-\nabla E_{i,j}^l(t)}\right)^2\right]^{\frac{1}{2}}\left[\sum_i\sum_j\sum_l\left(\Delta w_{i,j}^l(t-1)-\overline{\Delta w_{i,j}^l(t-1)}\right)^2\right]^{\frac{1}{2}}} \tag{2.30}$$

and

$$\overline{-\nabla E_{i,j}^l(t)}=-\frac{\sum_i\sum_j\sum_l \nabla E_{i,j}^l(t)}{n}$$

$$\overline{\Delta w_{i,j}^l(t-1)}=\frac{\sum_i\sum_j\sum_l \Delta w_{i,j}^l(t-1)}{n}$$

where t indexes the presentation number and n is the total number of weights. $\overline{-\nabla E_{i,j}^l(t)}$ is the mean value of negative error gradient with respect to $w_{i,j}^l$ in the layer l and $\overline{\Delta w_{i,j}^l(t-1)}$ is the mean value of previous weight change of weight $w_{i,j}^l$. The correlation coefficient measures the change in the direction of minimisation in a statistical way. From this correlation coefficient, three different conditions can be identified from this correlation coefficient.

A) When the correlation coefficient is near to one, there is almost no change in the direction of error minimisation and the change of weights is likely moving on the plateau. The learning rate can be increased to improve the convergence rate.

B) When the correlation coefficient is near to minus one, it implies an abrupt change in the direction of error minimisation, which is likely moving along the wall of the ravine. The learning rate should then be reduced to prevent oscillation across both sides of the ravine.

C) When there is no correlation between the negative gradient and previous weight update, the learning rate should be kept constant.

The following heuristic algorithm is suggested to alter the learning rate:

(i) $r(0)=0$ and set the value of $t=1$

(ii) **if** $r(t)>0$

if $r(t-1)>0$ **then** $\eta(t)=\eta(t-1)[1+dr(t)]$

else $\eta(t) = 1 + dr(t)$

if $r(t) < 0$

if $r(t-1) > 0$ **then** $\eta(t) = 1 + dr(t)$

else $\eta(t) = \eta(t-1)[1 + dr(t)]$

if $r(t) = 0$ **then** $\eta(t) = \eta(t-1)$

(iii) Set the value of $t = t + 1$, and repeat step (ii)

where d is a positive constants which determines how much the learning rate changes in each epoch. One of the significant features of this algorithm is the exponential increasing and decreasing of the learning rate. The learning rate can increase or decrease rapidly when the successive values of the correlation coefficient have the same sign. This feature enables the optimal learning rate to be found in a few learning iterations, and thus reduces the total output error rapidly. When the correlation coefficient changes sign, the algorithm can reset the excessive large value of learning rate and enhance the stability of this algorithm. As the learning rate is changed abruptly when the correlation coefficient changes sign, the algorithm is found more suitable to be used in the problems with binary target values. In these problems, the algorithm can drive the neurons to their target value quickly. Note that in a wide range of tests performed with this heuristic algorithm, it was found that a value of d equal to 0.5 was able to produce good results.

However, the convergence rate is not optimised with a fixed momentum coefficient α. The momentum term has an accelerating effect only when the $-\nabla E(t)$ and $\Delta W(t-1)$ have the same direction. For the fixed momentum coefficient, the momentum term may override the gradient term when the $-\nabla E(t)$ and $\Delta W(t-1)$ are in opposite directions. The momentum coefficient $\alpha(t)$ is suggested to be

$$\alpha(t) = \lambda(t)\eta(t)\frac{\|-\nabla E(t)\|_2}{\|\Delta W(t-1)\|_2} \tag{2.31}$$

where

$$\|-\nabla E(t)\|_2 = \left(\sum_i \sum_j \sum_l \left(-\nabla E_{i,j}^l(t)\right)^2\right)^{\frac{1}{2}}$$

$$\|\Delta W(t-1)\|_2 = \left(\sum_i \sum_j \sum_l \left(\Delta w_{i,j}^l (t-1) \right)^2 \right)^{\frac{1}{2}}$$

and

$$\lambda(t) = \begin{cases} \eta(t) & \eta(t) < 1 \\ 1 & \eta(t) \geq 1 \end{cases} \tag{2.32}$$

The final algorithm becomes

$$\begin{aligned} \Delta W(t) &= \eta(t)(-\nabla E(t)) + \alpha(t)\Delta W(t-1) \\ &= \eta(t)\|-\nabla E(t)\|_2 \left(-\nabla \hat{E}(t)\right) + \eta(t)\|-\nabla E(t)\|_2 \Delta \hat{W}(t-1) \\ &= \eta(t)\|-\nabla E(t)\|_2 \left(-\nabla \hat{E}(t) + \lambda(t)\Delta \hat{W}(t-1)\right) \end{aligned} \tag{2.33}$$

$-\nabla \hat{E}(t)$ and $\Delta \hat{W}(t-1)$ are the unit vectors of $-\nabla E(t)$ and $\Delta W(t-1)$, respectively. When $-\nabla \hat{E}(t)$ and $\Delta \hat{W}(t-1)$ are in opposite directions, $\lambda(t)$ is less than or equal to one, and hence the momentum term does not override the gradient term. When $-\nabla \hat{E}(t)$ and $\Delta \hat{W}(t-1)$ have the same direction, $\lambda(t)$ is greater than 1.0, and enhance the accelerating effect of the momentum term. The dramatic increase in learning rate may lead the network to get stuck in a local minimum in some neural network problems. This phenomenon can be avoided by the following strategy. If the present Root Mean Squares (RMS) error is greater than the previous one by 0.1, the last step is cancelled and $\eta(t)$ is reduced by half. This strategy gives a small preference to learning rate reduction and enhances the robustness of the training process.

2.3.2. Least Squares Based Training Algorithm

The algorithm decomposes each neuron into a linear part and a nonlinear part, i.e., the dot product of inputs and weights, and the activation function. The weights of neural networks are firstly initialized by a random generator. Given the last layer of weights and the inverses of activation function of the desired outputs of neurons in the output layer, the optimal outputs of the last hidden layer can be evaluated by a linear least squares method. As the solutions may be out of the range of the activation function, the solutions are scaled back into the range of the activation function by multiplying the solutions with a transformation matrix. After evaluating the desired outputs of the last hidden layer, the optimal weights between the last hidden layer and its preceding layer are found again by the least squares method. Subsequently, the optimal outputs of the layer preceding the last hidden layer are determined. The process is repeated until the weights between the input layer and first hidden layer are determined. The outputs of the last hidden layer are evaluated by propagating the training patterns through the network using the evaluated optimal weights. Subsequently, the weights between the last hidden layer and the output layer are evaluated. All the above procedures complete one training iteration and the process is repeated until the required accuracy is reached. The detailed mathematical description is described in the following.

A multilayer neural network with L fully interconnected layers is considered. Layer l consists of neurons ($l = 1, \ldots, L-1$) in which the last neuron is a bias node with a constant output of 1. If there are P patterns for network training, all given inputs can be represented by a matrix A^1 with P rows and $n_1 + 1$ columns. All entries of the last column of the matrix A^1 are constant 1. Similarly, the target can be represented by a matrix T^L with P rows and n_L columns. The weights between neurons in the layers l and $l+1$ form a matrix W^l with entries $w_{i,j}^l$ ($i = 1, \ldots, n_l + 1$, $j = 1, \ldots, n_{l+1}$). Entry $w_{i,j}^l$ connects neuron i of layer l with neuron j of layer $l+1$.

The output of all hidden layers and the output layer are obtained by propagating the training patterns through the network. Let us define the matrix

$$O^l = A^l W^l \tag{2.34}$$

To facilitate the description of our algorithm, we also define

$$A^l = \begin{bmatrix} A_p^l & B \end{bmatrix}, \text{ and}$$

$$W^l = \begin{bmatrix} U^l \\ V^l \end{bmatrix} \tag{2.35}$$

where B represents the outputs contributed by the bias node of layer l. B is a column vector with P entries, and every entry has the value of 1. A_p^l represents all the outputs of neurons in the layer l except the bias node. V^l represents the weights of connections between the bias node of layer l and neurons of layer $l+1$. V^l forms a row vector with dimension n_{l+1}. Therefore, O^l can also be expressed as

$$O^l = \begin{bmatrix} A_p^l & B \end{bmatrix} \begin{bmatrix} U^l \\ V^l \end{bmatrix} \tag{2.36}$$

The entries of A^{l+1} for all layers (i.e., $l = 1,\dots,L-1$) are evaluated as follows:

$$a_{p,j}^{l+1} = f(o_{p,j}^l) \qquad p = 1,\dots,P \quad \text{and} \quad j = 1,2,\dots,n_{l+1} \tag{2.37}$$

where $f(x)$ is the activation function.

Learning is achieved by adjusting the weights such that A^L is as close as possible or equal to T^L. Let us introduce a matrix S^{L-1} with entries

$$s_{i.j}^{L-1} = f^{-1}(t_{i,j}^L) \tag{2.38}$$

where $t_{i,j}^L$ are the entries of T^L.

We can reformulate the task of learning as least squares problem:

$$\min \left\| A^{L-1} W^{L-1} - S^{L-1} \right\|_2 \quad \text{or}$$

$$\min \left\| \begin{bmatrix} A_p^{L-1} & B \end{bmatrix} \begin{bmatrix} U^{L-1} \\ V^{L-1} \end{bmatrix} - S^{L-1} \right\|_2 \tag{2.39}$$

This linear least squares problem can be solved by QR factorization using Householder transforms or by Singular Value Decomposition (SVD). SVD is more robust when the number of equations is smaller than the number of unknown variables, however, it is not the case in training of neural networks. In addition, SVD is more computational complex than QR factorisation; therefore QR factorisation is used here. The Euclidean norm in Eq. (2.33) can be minimized by either adjusting

A_p^{L-1} or W^{L-1}. In this training algorithm, the optimal A_p^{L-1} is evaluated first. This problem is equivalent to find the solution of A_p^{L-1} in

$$\left\| A_p^{L-1} * U^{L-1} - \left(S^{L-1} - B * V^{L-1} \right) \right\|_2 = \text{minimal} \tag{2.40}$$

The optimal output of hidden layer is now the new target values for lower layers; i.e.,

$$T^{L-1} = A_p^{L-1} \tag{2.41}$$

With the optimal output of hidden layer, the weights of the lower layer can be adjusted to reach T^{L-1} as close as possible, but the entries in the matrix T^{L-1} may have values outside the range of the activation function. In order to transform T^{L-1} to a matrix with entries values inside the range, a transformation matrix C^{L-1} with order $\left(n_{L-1}+1\right)*\left(n_{L-1}+1\right)$ is constructed such that

$$\left[T_{trans}^{L-1} \quad B \right] = \left[T^{L-1} \quad B \right] * C^{L-1} \tag{2.42}$$

If the activation function is the standard sigmoid function with the range between 0 and 1, i.e.

$$f(x) = \frac{1}{1+\exp(-x)} \tag{2.43}$$

the elements of C^{L-1} is defined as

$$c_{d,d}^{L-1} = 1.0, \quad c_{d,k}^{L-1} = -\frac{\alpha_k - 1.25 * \max(0.5, \beta_k - \alpha_k, \alpha_k - \gamma_k)}{2.5 * \max(0.5, \beta_k - \alpha_k, \alpha_k - \gamma_k)}$$

$$c_{k,k}^{L-1} = \frac{1}{2.5 * \max(0.5, \beta_k - \alpha_k, \alpha_k - \gamma_k)} \tag{2.44}$$

where $d = n_{L-1} + 1$ and

$$\alpha_k = \sum_{j=1}^{P} \frac{t_{j,k}^{L-1}}{P}$$

$$\beta_k = \max_{j=1}^{P} t_{j,k}^{L-1}$$

$$\gamma_k = \min_{j=1}^{P} t_{j,k}^{L-1}, \quad \mathrm{k} = 1, \ldots, n_{L-1} \tag{2.45}$$

Values of all other entries of C^{L-1} are zeros. The following function is actually implemented by Eq. (2.36) with the values of $c_{i,j}^{L-1}$ defined as above.

$$t_{i,j(trans)}^{L-1} = \frac{t_{i,j}^{L-1} - \left(\alpha_j - 1.25 * \max\left(0.5, \beta_j - \alpha_j, \alpha_j - \gamma_j\right)\right)}{2.5 * \max\left(0.5, \beta_j - \alpha_j, \alpha_j - \gamma_j\right)}$$

$$= \frac{t_{i,j}^{L-1} - \alpha_j}{2.5 * \max\left(0.5, \beta_j - \alpha_j, \alpha_j - \gamma_j\right)} + 0.5 \tag{2.46}$$

From the Eq. (2.46), it can be proved easily that the average output of each neuron is 0.5. This equation summarizes three different mapping methods into one equation and which mapping method is used depends on the value of $\max\left(0.5, \beta_j - \alpha_j, \alpha_j - \gamma_j\right)$. When the values of $\beta_j - \alpha_j$ and $\alpha_j - \gamma_j$ are smaller than the half range of the activation function; i.e., smaller than 0.5, the Eq. (2.46) is simplified to

$$t_{i,j(trans)}^{L-1} = 0.4 * \left(t_{i,j}^{L-1} - \alpha_j\right) + 0.5 \tag{2.47}$$

It is noticed that $t_{i,j}^{L-1} - \alpha_j$ has a value in the range between $-\left(\gamma_j - \alpha_j\right)$ and $\beta_j - \alpha_j$. As $-\left(\gamma_j - \alpha_j\right)$ is greater than -0.5 and $\beta_j - \alpha_j$ is smaller than 0.5, $t_{i,j(trans)}^{L\ 1}$ is always between 0.1 and 0.9. When the value of $\max\left(0.5, \beta_j - \alpha_j, \alpha_j - \gamma_j\right)$ is equal to $\beta_j - \alpha_j$, the Eq. (2.46) is simplified to

$$t_{i,j(trans)}^{L-1} = 0.4 * \frac{t_{i,j}^{L-1} - \alpha_j}{\beta_j - \alpha_j} + 0.5 \tag{2.48}$$

In this case, the maximum value of $t_{i,j(trans)}^{L-1}$ is equal to 0.9 and the minimum value of $t_{i,j(trans)}^{L-1}$ is greater than 0.1. When the value of $\max\left(0.5, \beta_j - \alpha_j, \alpha_j - \gamma_j\right)$ is equal to $\alpha_j - \gamma_j$, the Eq. (2.47) is simplified to

$$t_{i,j(trans)}^{L-1} = 0.4 * \frac{t_{i,j}^{L-1} - \alpha_j}{\alpha_j - \gamma_j} + 0.5 \tag{2.49}$$

In this case, the minimum value of $t_{i,j(trans)}^{L-1}$ is equal to 0.1 and the maximum value of $t_{i,j(trans)}^{L-1}$ is smaller than 0.9. Therefore, the entries of S^{L-2} will not have very large negative and positive values. If the activation function used is

$$f(x) = \frac{1 - \exp(-x)}{1 + \exp(-x)} \tag{2.50}$$

the elements of C^{L-1} is defined as

$$c_{d,d}^{L-1} = 1.0, \quad c_{d,k}^{L-1} = -\frac{\alpha_k}{1.25 * \max(1.0, \beta_k - \alpha_k, \alpha_k - \gamma_k)}$$

$$c_{k,k}^{L-1} = \frac{1}{1.25 * \max(1.0, \beta_k - \alpha_k, \alpha_k - \gamma_k)} \tag{2.51}$$

Again values of all other entries of C^{L-1} are zeros. The following function is actually implemented by Eq. (2.42) with the values of $c_{i,j}^{L-1}$ defined by Eqs. (2.46-2.47).

$$t_{i,j(trans)}^{L-1} = \frac{t_{i,j}^{L-1} - \alpha_j}{1.25 * \max(1.0, \beta_j - \alpha_j, \alpha_j - \gamma_j)} \tag{2.52}$$

Using the similar argument as before, it can be shown that all entries of T_{trans}^{L-1} are between -0.8 and 0.8, and the average output of each neuron is zero. Therefore, the entries of S^{L-2} will not have very large negative and positive values. The next step is to train the lower layers of the network to produce T_{trans}^{L-1}.

If the network has three layers, by applying Eqs. (2.38) and (2.39) with L and T substituted by 2 and T_{trans} respectively, the optimal weights W^1 between the input and hidden layer can be determined by the least squares method. Propagating the training patterns through the network using the determined optimal weights W^1, updated hidden layer outputs A^2 can be evaluated accordingly. The optimal weights W^{L-1} between the hidden and output layers can be evaluated by applying Eq. (2.39) with L=3 and using the updated A^2. This process is repeated until the specified error level is reached.

For a network with more than one hidden layer, the weight between the hidden layers $l-1$ and l (for $l=3,\ldots,L-1$) are evaluated as follows. We start with $l=L-1$ after evaluating T_{trans}^{L-1} as stated in the previous section. The S^{l-1} is evaluated as in Eq. (2.38) with L replaced by l. Afterwards, W^{l-1} is first evaluated by the least squares method to minimize the Euclidean norm as in Eq. (2.39). After finding the optimal weights between layers $l-1$ and l, the A_p^{l-1} is found by Eq. (2.38) with L replaced by l. To transform T^{l-1} to a matrix with entry values inside the range of the activation function, a transformation matrix C^{l-1} is constructed as in Eq. (2.44) or Eq. (2.51), which is dependent on the activation function used. To maintain the least squares solution during forward propagating, an inverse matrix of C^{l-1} has to be found such that

$$W_{trans}^{l-1} = \left[C^{l-1}\right]^{-1} * W^{l-1} \tag{2.53}$$

Using the Eq. (2.53), the following equality holds when the input patterns propagate through the network.

$$\begin{aligned} \left[T_{trans}^{l-1} \quad B\right] * W_{trans}^{l-1} &= \left[T^{l-1} \quad B\right] * C^{l-1} * \left[C^{l-1}\right]^{-1} * W^{l-1} \\ &= \left[T^{l-1} \quad B\right] * W^{l-1} \\ &= A^{l-1} * W^{l-1} \end{aligned} \tag{2.54}$$

Therefore, the outputs of the network remain the same after the introduction of the transformation matrix and the least squares solution is thus valid. The entries of inverse $\left[C^{l-1}\right]^{-1}$ can be evaluated from those of C^{l-1} by the following equations:

$$\left(c_{d,k}^{l-1}\right)^{-1} = -\frac{c_{d,k}^{l-1}}{c_{k,k}^{l-1}}, \qquad \mathrm{k} = 1, \ldots, n_{l-1}$$

$$\left(c_{k,k}^{l-1}\right)^{-1} = \frac{1}{c_{k,k}^{l-1}}, \qquad \mathrm{k} = 1, \ldots, d \tag{2.55}$$

where $d = n_{l-1} + 1$.

The weights between hidden layers $l-1$ and l and optimal target T_{trans}^{l-1} are evaluated successively as stated in the previous paragraph until $l = 3$. The optimal weights W^1 between the input layer and hidden layer are then evaluated by solving the least squares problems as in Eq. (2.39). By propagating the training patterns through the network using the optimal weights, A^{L-1} is evaluated. Subsequently, the optimal weights W^{L-1} between the last hidden layer and the output layer are evaluated again by solving Eq. (2.39). This process is repeated until the specified error level is reached.

The LSB algorithm can be summarized as follows:

1. (i) The weights of the network are randomly initialized with values between -0.5 to 0.5.
   (ii) Evaluate S^{L-1} in Eq. (2.38).
2. Propagating all given patterns through the network so that the matrices A^l, $l = 2,\ldots,L$, can be successively evaluated. If error norm between A^L and T^L is smaller than the specified value, the training is completed.
3. (i) Evaluate T^{L-1} by solving the linear least squares problem in Eqs. (2.40) and (2.41).
   (ii) Evaluate the transformed matrix T_{trans}^{L-1} by applying Eq. (2.42) so that all the values of its entries are within the range of activation function.
   (iii) Set $T^{L-1} = T_{trans}^{L-1}$
4. For $l = L-1,\ldots,3$
   (i) Evaluate S^{l-1} as in Eq. (2.38).
   (ii) Evaluate a new set of weights W^l by solving the least squares problem as in Eq. (2.39).
   (iii) Evaluate T^{l-1} by solving the linear least squares problem as in Eqs. (2.40) and (2.41).
   (iv) Evaluate the transformed matrix T_{trans}^{l-1} by applying Eq. (2.42).
   (v) Evaluate the transformed weight matrix W_{trans}^{l-1} by applying Eq. (2.53).
   (vi) Set $T^{l-1} = T_{trans}^{l-1}$

5. (i) Evaluate S^1 as in Eq. (2.38).
   (ii) Evaluate a new set of weights W^1 by solving the least squares problem as in Eq. (2.39).
6. Evaluate the updated A^l ($l = 2,\ldots, L-1$) by propagating the input patterns through the network with the new weights.
7. Evaluate a new set of weights W^{L-1} by solving the least squares problem as in Eq. (2.39).
8. Evaluate the updated A^L with the new weights W^{L-1} obtained in step 7. If error norm between A^L and T^L is smaller than the specified value, the training is completed.
9. Go to step 3.

2.3.3. Extended Least Squares Based Algorithm

An extended least squares based algorithm for feedforward networks is described. This algorithm combines the features of the extended backpropagation algorithm described in 2.3.1 and the features of the pure least squares algorithm described in 2.3.2. The weights connecting the last hidden and the output layers are firstly evaluated by a least squares algorithm. The weights between the input and the hidden layers are then evaluated using the extended backpropagation algorithm. This arrangement eliminates the stalling problem experienced by the pure least squares type algorithms and still maintains the characteristic of fast convergence.

A multilayer neural network with L fully interconnected layers is considered to be same as Section 2.3.2 described. Learning is achieved by adjusting the weights such that A^L is as close as possible or equal to T^L so that the mean squared error E is minimized, where E is defined as

$$E = \frac{1}{2P} \sum_{p=1,\ldots,P} \sum_{j=1,\ldots,n_L} \left(a_{p,j}^L - t_{p,j}^L \right)^2 \tag{2.56}$$

In this learning algorithm, the weights between the last hidden layer and the output layer are evaluated by a pure least squares algorithm; the weights between the input and the hidden layers, the weights between hidden layers are evaluated by a modified gradient descent algorithm. The problem of determining the W^{L-1} optimally can be formulated as follows:

$$\min \left\| A^{L-1} W^{L-1} - T^{L} \right\|_2 \text{ with respect to } W^{L-1} \tag{2.57}$$

This linear least squares problem can be solved by using QR factorization together with Householder transforms or Singular Value Decomposition (SVD). QR factorization using Householder transforms was implemented because it has less computational complexity than SVD. After the optimal weights W^{L-1} are found, the new network output A^L are evaluated. To determine the appropriate weights change in the preceding layer, the remaining error is backpropagated to the preceding layer of the neural network. After the gradient information is obtained, the appropriate learning rate and momentum coefficient for each layer are determined in accordance with the correlation between the negative error gradient and the previous weight update of that layer (Yam & Chow, 1993). The correlation coefficient between the negative gradient and the last weight update for the layer l is given by

$$r^l(t) = \frac{\sum_i \sum_j \left(-\nabla E_{i,j}^l(t) - \overline{-\nabla E_{i,j}^l(t)}\right)\left(\Delta w_{i,j}^l(t-1) - \overline{\Delta w_{i,j}^l(t-1)}\right)}{\left[\sum_i \sum_j \left(-\nabla E_{i,j}^l(t) - \overline{-\nabla E_{i,j}^l(t)}\right)^2\right]^{\frac{1}{2}} \left[\sum_i \sum_j \left(\Delta w_{i,j}^l(t-1) - \overline{\Delta w_{i,j}^l(t-1)}\right)^2\right]^{\frac{1}{2}}} \tag{2.58}$$

where t indexes the presentation number, $-\nabla E_{i,j}^l(t)$ is the negative error gradient with respect to $w_{i,j}^l$ in the layer l and $\Delta w_{i,j}^l(t-1)$ is the previous weight change of weight $w_{i,j}^l$. $\overline{-\nabla E_{i,j}^l(t)}$ and $\overline{\Delta w_{i,j}^l(t)}$ are the mean values of the negative error gradients and the weight changes in the layer l respectively. The correlation coefficient is used again for measuring the change in the direction of minimisation in a statistical way. From this correlation coefficient, three different conditions can be identified.

A) When the correlation coefficient is near to one, there is almost no change in the direction of local error minimization and the change of weights is likely moving on the plateau. The learning rate can be increased to improve the convergence rate.

B) When the correlation coefficient is near to minus one, it implies an abrupt change in the direction of local error minimization, which is likely moving along the wall of the ravine. The learning rate should then be reduced to prevent oscillation across both sides of the ravine.

C) When there is no correlation between the negative gradient and the previous weight change, the learning rate should be kept constant.

According to these three conditions, the following heuristic algorithm is proposed to adjust the learning rate:

$$\eta^l(t) = \eta^l(t-1)\left(1 + 0.5r^l(t)\right) \tag{2.59}$$

The heuristic algorithm is a simplified version of the learning rate changing rules. The learning rate is changed abruptly when the correlation coefficient changes sign. The abrupt change in the learning rate often drives the neurons to their extreme regions where the targets of binary problem lie, and thus increases the rate of convergence. For the problem with continuous target values, the abrupt change in the learning rate may drive the neurons to overshoot their target values and cause oscillation.

The convergence rate is not optimized with a fixed momentum coefficient. The momentum term has an accelerating effect only when the $-\nabla E_{i,j}^l$ and $\Delta w_{i,j}^l$ have the same directions. For the fixed momentum coefficient, the momentum term may override the negative gradient term when the $-\nabla E_{i,j}^l$ and $\Delta w_{i,j}^l$ are in the opposite directions. The momentum coefficient $\alpha^l(t)$ at the t-th iteration is determined as follows:

$$\alpha^l(t) = \lambda^l(t)\eta^l(t)\frac{\left\|-\nabla E^l(t)\right\|_2}{\left\|\Delta w^l(t-1)\right\|_2} \tag{2.60}$$

where

$$\left\|-\nabla E^l(t)\right\|_2 = \left(\sum_i \sum_j \left(-\nabla E_{i,j}^l(t)\right)^2\right)^{\frac{1}{2}}$$

and

$$\left\|\Delta w^l(t-1)\right\|_2 = \left(\sum_i \sum_j \left(\Delta w^l(t-1)\right)^2\right)^{\frac{1}{2}} \tag{2.61}$$

$$\lambda^l(t) = \begin{cases} \eta^l(t) & \eta^l(t) < 1 \\ 1 & \eta^l(t) \geq 1 \end{cases} \tag{2.62}$$

As $\lambda^l(t)$ is always less than or equal to 1, the momentum term will not override the negative gradient term.

After evaluating the $\eta^l(t)$ and $\alpha^l(t)$ for layers $l = L-2,\ldots,1$, the new weights are determined. After that, the new network output and error are evaluated. One epoch is completed. As a fast increase in the learning rate may drive the neurons to their saturation region in some neural network problems, the following strategy is used to improve the stability of the algorithm. If the present Root Mean Squares Error (RMSE) is greater than the previous one by 0.1, the last weight change is canceled and $\eta^l(t)$ is reduced by half. This strategy gives a small preference to the learning rate reduction and enhances the robustness of the training process. The process is repeated until the network error reaches a specified error level.

The algorithm can be summarized as follows:

1. Generate initial weights for each layer using a pseudo-random generator.
2. Propagate all given patterns through the network so that the matrices A^l, $l = 2,\ldots,L$, can be successively evaluated. If the error norm between A^L and T^L is smaller than the specified value, the training is completed.
3. Evaluate a new set of weights W^{L-1} by solving the least squares problem.
4. Compute a new A^L from the A^{L-1} and the new weights W^{L-1}. If the error norm between A^L and T^L is smaller than the specified value, the training completes.
5. Compute the gradient for layers $l = L-2,\ldots,1$ using the new A^L and the new weights W^{L-1}.
6. For $l = L-2,\ldots,1$:
   (i) Evaluate the learning rate η^l and the momentum coefficient by applying Eqs. (2.58)-(2.62).
   (ii) Evaluate the weight change $\Delta w_{i,j}^l$.
   (iii) Evaluate the new weights W^l.

7. Compute a new A^L from A^1, W^1,..., W^l, W^{l+1}..., and W^{L-1}. If error norm between A^L and T^L is smaller than the specified value, the training completes.
8. If the RMS error is larger than the previous RMS error by 0.1, the weight changes for all layers are cancelled; the learning rate is reduced by half, and go to 6 (ii) again.
9. Continue with 2.

2.4. Initialization Consideration

There are several approaches to estimate optimal values for the initial weights so that the number of training iterations is reduced. It is quite obvious that the training session can be shortened when the starting point of the optimization is very close to the true minimum.

Shepanski (1988) regards the training of a multilayer feedforward network as an optimal estimation problem. The optimal set of weights is determined using a least squares criterion, employing a standard pseudoinverse matrix technique. This technique has been applied previously on a single layer network. Shepanski has extended the pseudoinverse method to multilayer networks. In a task of restoring a data bit stream, this method can reduce the initial error dramatically. Nguyen and Widrow speed up the training process by setting the initial weights of the hidden layer so that each hidden node is assigned to approximate a portion of the range of the desired function at the start of training (Nguyen & Widrow, 1990). By approximating the activation function with piece-wise linear segments, the weights can be evaluated. Next, the thresholds of neurons are selected by assuming the input variables are kept between -1 and 1. Nguyen and Widrow applied this method to initialize weights of numerous problems including the "Truck-Backer-Upper" problem. The results show that this method can greatly reduce the training time. Osowski (1993) following the idea of Nguyen and Widrow, developed another approach to select the initial values of weights of single input and single output three-layer feedforward neural networks. Osowski also approximates the activation function with piece-wise linear segments; however, he suggests an explicit method to determine the number of hidden neurons and uses the information of the desired function $y=f(x)$ to determine the initial weights. Firstly, the whole region of x of the measured (x,y) curve is split into sections containing only one negative or positive slope. After that, a set of neurons in the hidden layer is chosen equal in number to the number of sections determined. Each section of the curve is associated with one neuron in the hidden layer. The weights and the bias of each neuron are determined so that this neuron enters its active (middle) region of the proper section. Finally, the weights to the output layer are evaluated using the change of y in the corresponding section, and the bias of the output neuron is set to zero. In the example given in Osowski's paper, the optimum values of weights obtained by training the network using the BP algorithm are very close to the initial weights suggested by Osowski's algorithm. Hisashi Shimodaira (Shimodaira, 1994) proposes another method called OIVS to determine the distribution of initial values of the weights and the length of the weight vector so that the outputs of the neurons are in the active region in which the derivative of the sigmoid function has a large value. Shimodaira considers the input and output values of a neuron are located inside a unit hypercube (for the unipolar sigmoid function). By assuming that the activation region width in a neuron is larger than the length of the diagonal line of a unit hypercube and the center of the activation region coincides with the center of the unit hypercube, the weight value distribution and the magnitude of a weight value can be determined. In comparison with the BP algorithm, the OIVS method can greatly reduce the number of non-convergence cases in the training of the XOR problem. In the random mapping problem, the mean number of iterations required by the OIVS method is 0.46 times that required by the conventional BP algorithm. Drago and Ridella (1992) propose a method called Statistically Controlled Activation Weight Initialization (SCAWI) to find the optimal initial weights. The aim of the SCAWI method is to prevent neurons from saturation in the early stage by properly initializing the weights. Drago and Ridella determine the maximum magnitude of the weights by statistical analysis. They show that the maximum magnitude of the weights is a function of the paralysed neuron percentage (PNP), which is in turn related to the convergence speed. By determining the optimal range of PNP through computer simulations, the maximum magnitude of the weights can be obtained and the time needed to reach the required convergence criterion can be minimised. Denoeux and Lengellé (1993) suggest a method to initialize the weights in feedforward networks with one hidden layer. The proposed method relies on the use of reference patterns, or prototypes, and on a transformation that maps each vector in the original feature space onto a unit-length vector in a space with one additional dimension. Simulation results show that the method can reduce the training time, improve robustness against local minima and give better generalization.

In this section, three noniterative algorithms for determining the optimal initial weights of feedforward neural networks are suggested to enhance the convergence performance of MLP networks. All three algorithms are based on purely linear algebraic methods.

2.4.1. Weight Initialization Algorithm I

Considering a multilayer neural network with L fully interconnected layers is the same as describing in Section 2.3.2. The output of all hidden layers and the output layer are obtained by propagating the training patterns through the network. Let us define the matrix

$$O^l = A^l W^l \tag{2.63}$$

The entries of A^{l+1} for all hidden layers (i.e., $l = 1,\ldots,L-2$) are evaluated as follows:

$$a_{p,j}^{l+1} = f(o_{p,j}^{l}) \quad p = 1,\ldots,P \text{ and } j = 1,2,\ldots,n_{l+1} \tag{2.64}$$

$$a_{p,j}^{l+1} = 1.0 \qquad p = 1,\ldots,P \text{ and } j = n_{l+1} + 1 \tag{2.65}$$

where *f(x)* is the activation function. The activation function used here is the sigmoid function with range between 0 and 1:

$$f(x) = \frac{1}{1+\exp(-x)} \tag{2.66}$$

The entries of output layer A^L are evaluated as follows:

$$a_{p,j}^{L} = f(o_{p,j}^{L-1}) \quad p = 1,\ldots,P \text{ and } j = 1,2,\ldots,n_L \tag{2.67}$$

Learning is achieved by adjusting the weights such that A^L is as close as possible or equal to T^L. By introducing a matrix S with entries

$$s_{i.j}^{L-1} = f^{-1}(t_{i,j}^{L}) \tag{2.68}$$

where $t_{i,j}^{L}$ are the entries of T^L.

The weight initialization algorithm I (WIA I) starts by regarding the neural network being a single-layer perceptron network. The optimal weights for a single-layer perceptron can be evaluated by solving the following equation,

$$\text{minimize} \left\| A^1 W_L^1 - S^{L-1} \right\|_2 \tag{2.69}$$

This linear least squares problem can be solved by QR factorization using Householder reflections or Singular Value Decomposition (SVD). As SVD has better numerical stability, it is employed to solve the linear least squares problem. In the case of an overdetermined system, SVD produces a solution that is the best approximation in the least squares senses. In the case of an underdetermined system, SVD computes the minimal-norm solution.

The QR factorization of the optimal W_L^1 is then computed, i.e.,

$$W_L^1 = Q^1 R^1 \tag{2.70}$$

$$W_O^1 = Q^1 \tag{2.71}$$

W_O^1 contains the optimal weights connecting the input to the first hidden layer. In the QR factorization, we express an m-by-n matrix M as the product of an orthogonal matrix Q and an upper triangular matrix R, where $Q \in \Re^{m \times m}$ and $R \in \Re^{m \times n}$. QR factorization has property that when the columns of M are linearly independent, the first n columns of Q are an orthonormal basis for the columns of M (assuming $m \geq n$). As column vectors of W_O^1 are orthonormal, the product of input A^1 (scaled between 0.1 and 0.9) and W_O^1 seldom produces outputs that drive neurons to saturation. As the matrix W_L^1 has order $(n_1 + 1) \times n_2$, the matrix W_O^1 obtained by QR factorisation has order $(n_1 + 1) \times (n_1 + 1)$; therefore the number of neurons in the first hidden layer is restricted to be $n_2 = n_1 + 1$ when this algorithm is used.

After evaluating W_O^1, an updated A^2 is then evaluated using equations (2.63)-(2.65). After evaluating A^2, the network between the first hidden layer and the output layer is again considered as a single-layer perceptron network and the optimal weights W_L^2 is evaluated similarly as in Eq. (2.69). The initial values of weights W_O^2 are evaluated by QR factorization of W_L^2 in the same way as in Eq. (2.70). After evaluating the initial weights W_O^l between the l-th and the $l+1$-th layer, an updated A^{l+1} is calculated using eqs. (2.63)-(2.65). The network between the l+1-th layer and the output layer is solved as it is a linear discriminant, i.e.,

$$\text{minimize} \left\| A^{l+1} W_L^{l+1} - S^{L-1} \right\|_2 \tag{2.72}$$

The initial weights matrix W_O^{l+1} is set to equal to Q^{l+1} which is obtained from the QR factorization of W_L^{l+1}, i.e.,

$$W_L^{l+1} = Q^{l+1} R^{l+1} \tag{2.73}$$

$$W_O^{l+1} = Q^{l+1} \tag{2.74}$$

The process continues until $l = L-3$. After evaluating all the preceding layer of weights, an updated A^{L-1} is found. The optimal weight W_o^{L-1} is evaluated using the least squares method as in Eq. (2.72). All the initial weights are then found.

The number of neurons of layer $l+1$ is limited to $n_l + 1$ again because the matrix W_o^{l-1} obtained from the QR factorisation of the order of matrix W_L^l has order $(n_l + 1) \times (n_l + 1)$. In many cases, the architecture of a feedforward network must satisfy Kolmogorov's mapping theorem (1957) to ensure the function can be implemented by a three-layer feedforward neural network, i.e., $n_2 = 2n_1 + 1$. In these cases, an augmented matrix A_{aug}^{l+1} can be constructed from A^{l+1} by adding n_{l+1} column vectors with values 1.0, that is equivalent to adding n_{l+1} bias nodes to the layer $l+1$, i.e.,

$$A_{aug}^{l+1} = \left(A^{l+1} \quad \underbrace{\begin{matrix} 1 & 1 & \cdots & 1 \\ 1 & 1 & \cdots & 1 \\ \vdots & \vdots & \ddots & \vdots \\ 1 & 1 & \cdots & 1 \end{matrix}}_{n_{l+1}} \right) \tag{2.75}$$

By replacing A^{l+1} in Eq. (2.72) with A_{aug}^{l+1} and solving it, a matrix Q_{aug}^{l+1} of order $(2n_{l+1}+1) \times (2n_{l+1}+1)$ can be obtained by applying Eq. (2.73). As there is only one bias node in each layer of a standard feedforward neural network, the matrix W_O^{l+1} of order $(n_{l+1}+1) \times (2n_{l+1}+1)$ is constructed by copying the first n_{l+1} rows of Q_{aug}^{l+1} and setting the last row of W_O^{l+1} to be the sum of the last $n_{l+1}+1$ rows of Q_{aug}^{l+1}. With this arrangement, the product of W_O^{l+1} and R_{aug}^{l+1} still satisfies Eq. (2.72). Generally, the number of neurons in layer $l+2$ can be set to any value between $n_{l+1}+1$ and $2n_{l+1}+1$ by constructing an augmented matrix A_{aug}^{l+1} from A^{l+1}. However, this algorithm cannot be applied to the networks in which the number of hidden neurons is smaller than the number of input neurons plus one.

2.4.2. *Weight Initialization Algorithm II*

Although the WIA I can speed up the training process, this algorithm cannot be applied to the networks in which the number of hidden neurons is smaller than the number of input neurons plus one. To tackle this problem, the weights initialization algorithm II (WIA II) was proposed in (Yam & Chow, 1997). The rationale behind the second algorithm is to reduce the initial network error while preventing the network from getting stuck with the initial weights. The algorithm ensures the outputs of hidden layers are in the active region, i.e., the range in which the derivative of activation function has a large value. In the classical backpropagation algorithm, the weights are changed according to gradient descent direction of an error surface E,

$$E = \sum_{p=1}^{P} E_p \text{ and } E_p = \frac{1}{2}\sum_{j=1}^{n_L}\left(t_{p,j}^L - a_{p,j}^L\right)^2 \tag{2.76}$$

The weights are changed according to

$$\Delta w_{i,j}^l(t+1) = -\frac{\eta}{2P}\sum_{p=1}^{P}\frac{\partial E_p(t)}{\partial w_{i,j}^l(t)} \tag{2.77}$$

so that the error cost function E is minimized, where t labels the update epoch in the learning process, η is the learning rate. If the standard sigmoid function with the range between 0 and 1 as shown in Eq. (2.66) is used, the rule of changing the weights can be shown to be

$$\Delta w_{i,j}^l = \frac{\eta}{P}\sum_{p=1}^{P}\delta_{p,j}^l a_{p,i}^l \tag{2.78}$$

For the output layer, i.e. $l = L-1$

$$\delta_{p,j}^{L-1} = (t_{p,j}^L - a_{p,j}^L) f'(o_{p,j}^{L-1}) = (t_{p,j}^L - a_{p,j}^L) a_{p,i}^L (1 - a_{p,i}^L) \tag{2.79}$$

For the other layers, i.e. $l = 1 \ldots L-2$

$$\delta_{p,j}^l = f'(o_{p,j}^l)\sum_{k=1}^{n_{l+2}}\delta_{p,k}^{l+1} w_{k,j}^{l+1} = a_{p,i}^{l+1}(1 - a_{p,i}^{l+1})\sum_{k=1}^{n_{l+2}}\delta_{p,k}^{l+1} w_{k,j}^{l+1} \tag{2.80}$$

From Eqs. (2.78) to (2.80), it is noticed that the change of a weight depends on the outputs of neurons connected to it. When the output of neuron is 0 or 1, the derivative of the activation function evaluated at this value is zero. Therefore, there will be no weight change at all, even if there is a difference between the value of the target and the actual output. Instead of using a statistical method to evaluate the maximum magnitude of weights (Drago & Ridella, 1992), the outputs of hidden neurons are assigned with random numbers in the range between $1 - \bar{t}$ and $\bar{t}$. $\bar{t}$ is chosen that $\bar{t}(1 - \bar{t}) = 0.01$. Let us introduce a matrix S^l with entries

$$s_{i.j}^{l-1} = f^{-1}(t_{i,j}^l) \qquad l = 2 \ldots L-1 \tag{2.81}$$

where $t_{i,j}^l$ are the random numbers within the stated range and contained in the matrix T^l. The neurons of the first hidden layer are first assigned with random numbers in the specific range. Then the optimal weights for a single-layer perceptron can be evaluated by solving the following equation,

$$\text{minimize}\left\|A^1W^1 - S^1\right\|_2 \tag{2.82}$$

This linear least squares problem can be solved by QR factorization using Householder reflections. In the case of an overdetermined system, QR factorization produces a solution that is the best approximation in a least squares sense. In the case of an underdetermined system, QR factorization computes the minimal-norm solution. As the number of training patterns is always greater than the number of variables in W^1 in neural network applications, the system is always overdetermined and the least squares solution is obtained by solving the Eq. (2.82). After the weights W^1 are evaluated, the actual outputs of the first hidden layer A^2 can be evaluated by propagating the input patterns using the weights W^1. The outputs of hidden neurons A^2 are always in the range between $1-\bar{t}$ and $\bar{t}$ because the least squares solution W^1 is used. As the outputs of hidden neurons are not at the extreme state, the derivatives of the activation function at these values are not zeros and the network will never get stuck. After evaluating A^2, the outputs of the second hidden layer are assigned with T^3 in which entries have random values between $1-\bar{t}$ and $\bar{t}$. By solving the linear least squares problem, the optimal weights W^2 can be evaluated. In general, to find the optimal weights W^l (for $l = 1 \ldots L-2$), we have first find all the previous weights and generate a matrix T^{l+1} using a random number generator. By propagating the input patterns through the network using the evaluated optimal weights, A^l can be found. The optimal weights W^l can be evaluated by solving the following equation:

$$\text{minimize}\left\|A^lW^l - S^l\right\|_2 \qquad l = 1 \ldots L-2 \tag{2.83}$$

To evaluate the last layer of weights W^{L-1}, we find A^{L-1} first as stated before; however, we need not generate the matrix T^L using a random number generator, instead we directly use the target values of the training patterns. The following equation is used to find the last layer of weights:

$$\text{minimize}\left\|A^{L-1}W^{L-1} - S^{L-1}\right\|_2 \tag{2.84}$$

The weight initialization process is then completed.

2.4.3. *Weight Initialization Algorithm III*

It was noticed that several networks initialized by the WIA II got stuck when they were further trained by the BP algorithm. A modification to the WIA II algorithm was proposed to tackle this problem. By the same argument stated in the second algorithm, the outputs of the hidden neurons should be within the active region, i.e. the range in which the derivative of activation function has a large value, so that the learning will not be hindered. Instead of assigning the outputs of hidden units and using the least squares method to evaluate the weights, the magnitudes of weights required to ensure that the outputs of hidden units are in the active region are derived as in the following.

To evaluate the required range of weights $w_{i,j}^l$ when the activation function shown in Eq. (2.66) is used, the following problem has to be solved:

$$1-\bar{t} \le a_{p,j}^{l+1} \le \bar{t} \text{ or } -\bar{s} \le o_{p,j}^l \le \bar{s} \tag{2.85}$$

where $\bar{s} = f^{-1}(\bar{t})$. The Eq. (2.85) can be simplified to be

$$\left(o_{p,j}^l\right)^2 \le \bar{s}^2 \text{ or } \left(\sum_{i=1}^{n_l+1} a_{p,i}^l w_{i,j}^l\right)^2 \le \bar{s}^2 \tag{2.86}$$

By Cauchy's inequality,

$$\left(\sum_{i=1}^{n_l+1} a_{p,i}^l w_{i,j}^l\right)^2 \le \sum_{i=1}^{n_l+1}\left(a_{p,i}^l\right)^2 \sum_{i=1}^{n_l+1}\left(w_{i,j}^l\right)^2 \tag{2.87}$$

In order to ensure the outputs of neurons are in the active range, the requirement for the range of weights is further constricted using the Eq. (2.87), i.e.

$$\sum_{i=1}^{n_l+1}\left(a_{p,i}^l\right)^2 \sum_{i=1}^{n_l+1}\left(w_{i,j}^l\right)^2 \le \bar{s}^2 \tag{2.88}$$

If n_l is a large number and the weights are values between $-\theta_p^l$ to θ_p^l with uniform probability distribution, the following equation is obtained:

$$(n_l+1)E\left[\left(w_{i,j}^l\right)^2\right] \le \frac{\bar{s}^2}{\sum_{i=1}^{n_l+1}\left(a_{p,i}^l\right)^2} \tag{2.89}$$

where $E\left[\left(w_{i,j}^l\right)^2\right]$ is the second moment of the weights at layer l, which can be shown to be

$$E\left[\left(w_{i,j}^l\right)^2\right] = \int_{-\theta_p^l}^{\theta_p^l} \frac{x^2}{2\theta_p^l} dx$$

$$= \frac{\left(\theta_p^l\right)^2}{3} \tag{2.90}$$

Therefore, the magnitude of weights θ_p^l for pattern P is chosen to be:

$$\theta_p^l = \bar{s}\sqrt{\frac{3}{(n_l+1)\sum_{i=1}^{n_l+1}\left(a_{p,i}^l\right)^2}} \tag{2.91}$$

For different input patterns, the values of θ_p^l are different. To make sure the outputs of hidden neurons are in the active region for all patterns, the minimum value of θ_p^l is chosen:

$$\theta^l = \min_{p=1,\cdots,P}\left(\theta_p^l\right) \tag{2.92}$$

The weights at layer l are then initialized by a random number generator with uniform distribution between $-\theta^l$ to θ^l.

The procedures of the third weights initialization are as follows:

(i) Evaluate θ^1 using the input training patterns $a_{p,i}^1 = x_{p,i}$.

(ii) The weights $w_{i,j}^1$ are initialized by a random number generator with uniform distribution between $-\theta^1$ to θ^1.

(iii) Evaluate $a_{p,i}^2$ by propagating the input patterns through the network using $w_{i,j}^1$.

(iv) For $l = 2,\ldots,L-2$

a) Evaluate θ^l using the outputs of layer l, i.e. $a_{p,i}^l$.
b) The weights $w_{i,j}^l$ are initialized by a random number generator with uniform distribution between $-\theta^l$ to θ^l.
c) Evaluate $a_{p,i}^{l+1}$ by propagating the outputs of $a_{p,i}^l$ through the network using $w_{i,j}^l$.

(v) After finding $a_{p,i}^{L-1}$ or A^{L-1}, we can find the last layer of weights W^{L-1} using Eq. (2.86) as stated in the second weights initialization algorithm.

2.5. Global Learning Algorithms

Another fundamental limitation of the error Backpropagation algorithm is dominated by the susceptibility to local minima during learning process. In the past decade, many researchers have proposed different types of algorithms which can provide a significant reduction in the total number of iterations by exploiting the fast convergence characteristics, but getting stuck in the local minima is still inevitable for many cases. Convergence stalling is yielded especially in applying to the complicated problems.

Global optimization methods (Horst & Pardalos, 1995) were, recently, employed instead of the conventional gradient descent to avoid the convergence trapped in undesired local minima. For instance, Random Search (RS) method (Baba, 1989) is a stochastic optimization method, which could be successfully utilized to find the global minimum in the error surface. Simulated Annealing (SA) (Kirkpatrick et al., 1983) is another stochastic type global minimization method to allow any down-hill and up-hill movement for escaping from the local minima which results the search convergence to a global minimum at the end. Evolutionary Programming (Fogel, 1995) or Genetic Algorithm (Goldberg, 1985, Holland, 1975), based upon the computational model of evolution, is another mechanism for the global optimization to determine connectivity in the feedforward networks. All of these methods have been applied to complex and multi-model minimization problems with both discrete and continuous variables. Moreover, Battiti and Tecchiolli (1995) proposed a heuristic scheme of the reactive Tabu search, based on a "modified local search" component complemented with a meta-strategy, for the neural networks learning. Shang and Wah (1996) developed a trajectory-based non-linear optimization method. It relies on an external force to pull a search out of the local minima. Their global optimization method is capable of providing excellent results by applying in the pattern recognition and classification problems. However, the algorithm is extremely computational complex because the evaluation of the trace function, that leads the trajectory out of the local minima, requires highly computation for the solution of an ordinary differential equation. Another learning scheme for achieving global optimization is based on the hybrid local and global optimizations method. This method was operated by making use of local search to optimize the weights between the output and hidden layers and use global search to optimize the other weights at the other layers. One method was proposed in (Cho & Chow, 1999) which is based on the hybrid of Least Squares method and penalized optimization method. The idea of this penalized optimization is defined by superimposing a discontinuous type function under the weight space domain while the learning process is getting stuck. This study included the use of Gaussian function, Rayleigh function and Laplace function as the penalty-like functions. Despite the success of this approach in performing global learning, like all other algorithms, we are still unable to determine the minima distribution in error surfaces and unable to identify whether a minimum is a local one or global one.

2.5.1. Simulated Annealing Algorithm

The simulated annealing method has been introduced in 1983 by Kirkpatrick, Gellat and Vecchi, inspired by the annealing (cooling) process of crystals that reach the lowest energy, corresponding to the perfect crystal structure, if cooled sufficiently slowly. Simulated Annealing (SA) has found numerous applications in all branches of science and technology. There are three types of user-defined functions in the SA procedure: first, $p_p(P)$, describing the probability distribution of parameters; second, $p_E(\Delta E(P))$, the probability of accepting the new set of parameters as the current one, depending on the change of the error function; and third, $T(k)$, the schedule of changing the 'temperature' parameter T in some time steps t. Temperature T determines the scale of fluctuations allowed at a given time step. The Boltzmann annealing schedule is most frequently used because of the statistical mechanics roots of the method. It is defined by:

$$p_E(\Delta E(P)) = \frac{1}{1+\exp(\Delta E(P)/T)} \tag{2.93}$$

There are various proofs showing that, with the probability approaching one, for $T(t)$ shower than $T_0/\ln t$ a global minimum can be found by this procedure. For the p_p distribution gaussian form is frequently used:

$$p_p(\Delta P) = (2\pi T)^{N/2} \exp\left(-\Delta P^2/2T\right) \tag{2.94}$$

where ΔP is the vector defining change of parameters from the previous value. Another popular simulated annealing method, called Fast Annealing, is based on Cauchy distribution, defined by:

$$p_p(\Delta P) = \frac{T}{\left(\Delta P^2 + T^2\right)^{(N+1)/2}} \tag{2.95}$$

which assigns higher probabilities to larger changes of parameters. To save time, temperature is frequently reduced by a constant amount, leading to exponential schedule that does not guarantee that the global minimum is found. In many simulations in the initial stages of minimization will not allow to sample the minima; to avoid the time waste short sample runs with fast annealing schedule are recommended to determine good initial temperature. In later stages, when local minima are explored, shifting to gradient based or linear search techniques may significantly reduce cost of calculation.

2.5.2. Alopex Algorithm

A special form of simulated annealing is used in the Alopex algorithm (Unnikrishnan & Venugopal, 1994), since the result of this approach seem to be very good. Alpoex algorithm is based on a very simple idea which is competitive to the backpropagation. The weight w_{ji} is changed by a constant amount δ with probability defined by the sigmoidal factor, $p_{ji} = \sigma\left(\Delta w_{ji} \cdot \Delta E/T\right)$, where the weight change and the error change computed in the previous iteration is used. The annealing temperature is changed every epoch consisting of K steps, using the sum of error changes in the previous epoch:

$$T(n) = \frac{\delta}{K} \sum_{t=n-K}^{n-1} |\Delta E(t)| \tag{2.96}$$

For large temperature, probabilities of $\pm\delta$ are close to 0.5 and the weights are randomly changed until a large change of energy is detected (correlation between changes of weights and changes of error are large) and the temperature is reduced. During an iteration all weights are updated simultaneously. No assumptions are made about the structure of the network, the error measure being minimized or the transfer functions, no gradients are computed, the same algorithm may be used in feedforward as well as recurrent networks, and there is even some neurobiological plausibility of this algorithm. There are 3 parameters: the step-size δ, which is taken as 0.01-0.001 times the dynamic range of weights, the initial temperature, and the number of steps per epoch K=10-100. Obviously many improvements can be proposed, such as the variable K, fast and slow weights, different annealing schedules etc. Alopex may be quite easily used in connection with other global minimization methods, for example with genetic algorithms. One disadvantage of the Alopex algorithm seems to be that the weights are always updated and therefore saturate large positive or negative values. To prune the small weights and enable feature selection it is better to define conditions when they may vanish, for example by using penalized optimization described in the later section.

2.5.3. *Reactive Tabu Search*

The reactive tabu search is based on a simple idea. The search is started at a random point and the best elementary move is selected; cycles are avoided by keeping the trajectory of the search and discouraging the system from visiting the same regions again. In context of neural networks the values of the adaptive parameters P are kept with finite precision and the neighborhood $N(P)$ is defined by single-bit change operations. The error function $E(P)$ is therefore defined on a finite set of points. The best operation for which $E(P')$, $P' \in N(P)$ has the lowest value, is selected (even if the error grows) and the ties are broken in a random way. The inverses of most recent moves are prohibited to avoid cycles, hence the 'tabu' name for the method – regions already visited should be avoided. If there are too many possibilities only a restricted subset of moves are randomly sampled and the best one selected. The tabu is put on the moves, not on the values of P, and kept for a limited number of T time steps. The values of T should be large enough to avoid cycles and small enough to avoid over-constraining the search. The reactive tabu search optimizes the prohibition period T adjusting it to the local structure of the problem. This requires remembering the points P along the trajectory and counting how many times each point has been encountered. The reactive tabu search was used with very good result on a large number of combinatorial optimization problems. It has used to discriminate interesting events in High Energy Physics data, with the best results obtained for a one-bit representation of weights. The generalization levels reached 90% while in the standard BP they reached only 62%.

2.5.4. The NOVEL Algorithm

A hybrid, global/local trajectory based method, called NOVEL has been proposed for neural networks (Shang & Wah, 1996). This method was exploring the solution space, locating promising regions and using local search to locate promising minima. Trajectory $P(t)$ in the global search stage is defined by a differential equation:

$$\hat{P}(t) = A(\nabla_P M(P(t))) + B(T(t), P(t)) \tag{2.97}$$

where T is the trace function and A, B are in general non-linear functions. The first component allows local minima to attract the trajectories, and the second component allows to walk out from the local minima. In the simplest case used in the NOVEL algorithm A and B functions are constants:

$$\hat{P}(t) = -\mu_g \nabla_P M(P(t)) + \mu_t(T(t) - P(t)) \tag{2.98}$$

The trace function T should assure that all space is finally traversed; it may either partition the space into regions that are explored in details or make first coarse and fine searches. The differential equation is either solved in its original form by standard Ordinary Differential Equation (ODE) or in a discretized form as a differential equation:

$$P(t + \delta t) = P(t) + \delta t\left[-\mu_g \nabla_P M(P(t)) + \mu_t(T(t) - P(t))\right] \tag{2.99}$$

It was noted that ODE solutions are slightly better although discretized equations are faster to simulate. The method has been tested on the two-spiral problem, training 5 hidden units in 100 time steps, starting from zero weights. This is one of very hard problems for neural networks. This method claimed that finding a solution for 4 hidden units required a total of one week of Sun workstation running time.

2.5.5. *The Heuristic Hybrid Global Learning Algorithm*

A heuristic approach global learning algorithm, based on the hybrid of the Least Squares and the Penalized optimization methods, has been described in (Cho & Chow, 1999). For simplicity, we assume that the multilayer neural network has a single hidden layer. The weights connected between the output layer and the hidden layer are firstly determined using the Least Squares (LS) based method. Afterward, the weights connected between the hidden layer and the input layer are determined by the gradient descent optimization. When the learning process is stuck, the learning mechanism is switched to minimizing the new penalty approach through the weight matrix (**V**) at the output layer. In terms of its working mechanism, the weights (**V**) are estimated by the LS method whilst the weights (**W**) at the hidden layer are estimated by the penalty approach optimization.

As described in (Cho & Chow, 1999), let us define the global optimization problem to be considered as follows. Let $E(\mathbf{w}):\Re^n \to \Re$ be the cost function, which can be twice differentiable continuously, where **w** is a vector of n state variables or parameters. The objective is to find the global minimum solution, $\mathbf{w}_{gm}$ which minimizes $E(\mathbf{w})$,

$$E^* = E(\mathbf{w}_{gm}) = \min\{E(\mathbf{w}) | \mathbf{w} \in \Im\} \tag{2.100}$$

where $\Im$ is the domain of the state variables over which one seeks the global minimum and $\Im$ is assumed to be compact and connected. Further, we assume that every local minimum $\mathbf{w}_{lm}$ of $E(\mathbf{w})$ in $\Im$ satisfies the conditions

$$\frac{\partial E(\mathbf{w}_{lm})}{\partial \mathbf{w}} = 0 \tag{2.101}$$

$$\mathbf{y}^T\left(\frac{\partial^2 E(\mathbf{w}_{lm})}{\partial \mathbf{w}^2}\right)\mathbf{y} \geq 0, \qquad \forall \mathbf{y} = \Re^n \tag{2.102}$$

Based on the above conditions, we assume that the global minimum satisfies these local minima and that the global minimum does not occur beyond the boundary of $\Im$. A well-known penalized optimization method was introduced in (Zheng & Zhuang, 1992) stating that the problem (2.100) can be approximated by solution of associated penalized unconstrained problems in finite dimensional spaces as shown below:

$$\min\{E(\mathbf{w}) + \lambda E_{pen}(\mathbf{w})\} = \min\{E(\mathbf{w}) | \mathbf{w} \in \Im\} = E^* \tag{2.103}$$

where $E_{pen}(\mathbf{w})$ is a penalty function for the constraint set $\Im$ and λ is a real number that $\lambda > 0$. The penalized optimization technique provides an uphill force whenever the convergence is trapped in local minima. Thus, a discontinuous penalty function, defined from (Cho & Chow, 1999) for global optimization, should satisfy

$$E_{pen}(\mathbf{w}) = \begin{cases} 0, & \mathbf{w} \in \Im \\ g(\mathbf{w}) & \mathbf{w} \notin \Im \end{cases} \tag{2.104}$$

where $g(\mathbf{w})$ is a penalty-like function. Three types of the penalty-like function are chosen such as: Gaussian function, Rayleigh function and Laplace function.

For simplicity, the algorithm was considered to use a feedforward neural network with a single hidden layer network. Suppose the network has M input units, N hidden units and Q output units, and we assume that the activation function takes a form of $\sigma(x) = \dfrac{1}{(1 + \exp(-x))}$ in the hidden layer. The general form of the feedforward network can be represented as follows:

$$o_k = \sum_{j=1}^{N} v_{kj}\sigma\left(\sum_{i=1}^{M} w_{ji}x_i + b_j\right) + \theta_k, \qquad 1 \leq k \leq Q \tag{2.105}$$

where o_k and x_i denote network output and input values respectively. w_{ji} and b_j denote the synapses weights and bias term respectively connected between the hidden layer and the input layer which form elements of the weight matrix **W.** Similarly, v_{kj} and θ_k denote the synapses and bias respectively connected between the output layer and the hidden layer which form elements of the weight matrix **V**. Assume there are P patterns in the training set. For pattern $p = 1,2,\ldots,P$, let $\mathbf{t}_p = (t_{1p}, t_{2p}, \ldots, t_{Qp})^T$ and $\mathbf{o}_p = (o_{1p}, o_{2p}, \ldots, o_{Qp})^T$ denote the desired output vector and the true output vector of the network. $\mathbf{a}_p = (a_{1p}, a_{2p}, \ldots, a_{Np}, 1)^T$ denotes the vector of outputs at the hidden layer, where the entries $a_{jp} = \sigma\left(\sum_{i=1}^{M} w_{ji} x_{ip} + b_j\right)$. $\mathbf{x}_p = (x_{1p}, x_{2p}, \ldots, x_{Mp}, 1)^T$ denotes the vector of inputs of the network.

A total sum-squared-error function,

$$E_T(\mathbf{V}, \mathbf{W}) = \sum_{p=1}^{P} (\mathbf{t}_p - \mathbf{o}_p)^T (\mathbf{t}_p - \mathbf{o}_p) \tag{2.106}$$

is chosen as a cost function for the network optimisation. The goal of the learning algorithm is required to optimise the weights of the network by minimising the cost function such that all the derivatives of $E_T(\mathbf{V}, \mathbf{W})$ with respect to **V** are equalled to zero, so the optimal weight matrix **V** can be exactly computed by

$$\hat{\mathbf{V}} = \mathbf{T}\mathbf{A}^T (\mathbf{A}\mathbf{A}^T)^+ \tag{2.107}$$

where $\mathbf{T} = (\mathbf{t}_1, \mathbf{t}_2, \ldots, \mathbf{t}_P)$ and $\mathbf{A} = (\mathbf{a}_1, \mathbf{a}_2, \ldots, \mathbf{a}_P)$. "+" denotes an operation of pseudo-inverse of the matrix. Thus, the cost function $E_T(\mathbf{V}, \mathbf{W})$ can be reformulated in a form of

$$E_T(\mathbf{V}, \mathbf{W}) = E(\mathbf{V}, \mathbf{W}(\mathbf{V})) = E_{hid}(\mathbf{W}) \tag{2.108}$$

because the weight matrix **V** can be expressed as a matrix function of matrix **W**. There exists a new weight space $\Re^{M \times N}$ with lower dimension mapping into the original weight space. In other words, the minimisation of the new cost function $E_{hid}(\mathbf{W})$ is equivalent to minimising the error function over the whole weight space. In this study, the weights matrix, **W** can be iteratively estimated via minimising the cost function by the gradient descent optimisation.

As our aforementioned statements, the minimisation by the gradient descent optimisation suffers from the problem of local minima.

Therefore it is advisable to modify the new cost function, $E_{hid}(\mathbf{W})$ by including a penalty function to provide a search out of the local minima when the convergence gets stuck. In this paper, the penalty function is introduced to superimpose under the weight space domain. Three different types of our proposed penalty functions are given as follow,

$$E_{pen}(\mathbf{W},\lambda)=\lambda\sum_{j=1}^{N}\sum_{i=1}^{M}\exp\left(-\frac{\left|w_{ji}(n-1)-w_{ji}^{*}\right|^{2}}{\lambda^{2}}\right)$$

Gaussian type (2.109)

$$E_{pen}(\mathbf{W},\lambda)=\lambda\left|w_{ji}(n-1)-w_{ji}\right|\sum_{j=1}^{N}\sum_{i=1}^{M}\exp\left(-\frac{\left|w_{ji}(n-1)-w_{ji}\right|^{2}}{\lambda^{2}}\right)$$

Rayleigh type (2.110)

$$E_{pen}(\mathbf{W},\lambda)=\lambda\sum_{j=1}^{N}\sum_{i=1}^{M}\exp\left(-\frac{\left|w_{ji}(n-1)-w_{ji}\right|}{\lambda^{2}}\right)$$

Laplace type (2.111)

where $w_{ji}(n-1)$ is the weight in the specify weight space, $\Re^{M\times N}$ at the previous iteration. w_{ji} is the sub-optimal weight value which is assumed to be getting stuck in a local minimum. λ denotes the penalty factor which determines the influence of the penalty term against the original least squares cost function. This weighting factor is used to control the breadth of the search iteratively to force the trajectory out of the local minima. The correct choice and adaptation procedure of λ will be described later. In accordance with the penalty approach, the modified cost function for the proposed heuristic algorithm is defined as

$$E_{gbl}(\mathbf{W},\lambda)=E_{hid}(\mathbf{W})+E_{pen}(\mathbf{W},\lambda) \qquad (2.110)$$

Consequently, the updated equations for the weights connected between the hidden layer and the input layer is,

$$w_{ji}^{**}(n) = w_{ji}(n-1) + \eta \frac{\partial E_{gbl}(\mathbf{W},\lambda)}{\partial w_{ji}} \tag{2.112}$$

$$\frac{\partial E_{gbl}(\mathbf{W},\lambda)}{\partial w_{ji}} = -\frac{\partial E_{hid}(\mathbf{W})}{\partial w_{ji}} + \frac{\partial E_{pen}(\mathbf{W},\lambda)}{\partial w_{ji}} \tag{2.113}$$

where $w_{ji}^{**}(n)$ is an optimal weight in the specified weight space by minimising the modified cost function. η denotes a learning rate of the algorithm. $\frac{\partial E_{hid}(\mathbf{W})}{\partial w_{ji}}$ is the original gradient of $E_{hid}(\mathbf{W})$ with respect to w_{ji} and $\frac{\partial E_{pen}(\mathbf{W},\lambda)}{\partial w_{ji}}$ is the penalty term of the derivative of $E_{pen}(\mathbf{W},\lambda)$ with respect to w_{ji} using eqns. (2.109-2.110), so we define:

Gaussian term:

$$\frac{\partial E_{pen}(\mathbf{W},\lambda)}{\partial w_{ji}} = \frac{2}{\lambda}\left|w_{ji}(n-1) - w_{ji}\right| \exp\left(-\frac{\left|w_{ji}(n-1) - w_{ji}\right|^2}{\lambda^2}\right) \tag{2.114}$$

Rayleigh term:

$$\frac{\partial E_{pen}(\mathbf{W},\lambda)}{\partial w_{ji}} = \frac{2}{\lambda}\left|w_{ji}(n-1) - w_{ji}\right|^2 \exp\left(-\frac{\left|w_{ji}(n-1) - w_{ji}\right|^2}{\lambda^2}\right) \tag{2.115}$$

Laplace term:

$$\frac{\partial E_{pen}(\mathbf{W},\lambda)}{\partial w_{ji}} = \frac{1}{\lambda}\exp\left(-\frac{\left|w_{ji}(n-1) - w_{ji}\right|}{\lambda^2}\right) \tag{2.116}$$

as the proposed penalty terms. Based on the above formulations, these allow to descent the gradient of E_{hid} and to ascend the gradient of E_{pen}, that the classical gradient descent in the basin of attraction of a minimum and for escaping a local minimum entrapment were implemented in this algorithm.

In accordance with the eq. (2.114), the penalty factor λ determines the effect of the penalty term against the original negative gradient of $\frac{\partial E_{hid}(\mathbf{W})}{\partial w_{ji}}$. A correct selection and adaptation for λ are critical considerations for the entirely algorithm performance. The selection of λ is based on the following condition:

1. If λ is too small, large weight changes are possible because of the strong effect of the penalty term. In this case, the algorithm is unstable which may result an invalid solution.
2. If λ is too large, only very small weight changes are allowed for the trajectory to escape from the local minima because the penalty term becomes virtually redundancy.

The best choice of λ lies between these two extremes and with the condition of assuring the training convergence. The proposed algorithm is said to be convergent in the mean square if the mean-square value of the error vector, $\mathbf{e}_{gbl}(n)=\mathbf{t}_p-\mathbf{o}_p(n)$ approaches a constant value as the number of iterations n approaches infinity; that is:

$$E\left\{\mathbf{e}_{gbl}(n)^T\mathbf{e}_{gbl}(n)\right\}\rightarrow \text{constant} \qquad \text{as } n\rightarrow\infty \tag{2.117}$$

so, it implicit that,

$$\Delta E_{gbl}(n)=E_{gbl}(n+1)-E_{gbl}(n)\le 0 \qquad \text{as } n\rightarrow\infty \tag{2.118}$$

The modified cost function can be approximated by a first order Taylor series:

$$E_{gbl}(n+1)=E_{gbl}(n)+\left[\frac{\partial E_{gbl}(n)}{\partial \mathbf{w}_j}\right]^T\Delta\mathbf{w}_j(n) \tag{2.119}$$

where $\left[\frac{\partial E_{gbl}(n)}{\partial \mathbf{w}_j}\right]^T$ represents the gradient of the modified cost function with respect to the weight vector $\mathbf{w}_j$; $\Delta\mathbf{w}_j(n)$ represents the change in an arbitrary weight vector.

Based on the above expressions, suppose the Gaussian penalty term is a general case, the guideline of the selection of λ is defined as

$$0 < \lambda \le \left| 2 \left| \mathbf{w}_j(n-1) - \mathbf{w}_j \right| \left[\frac{\partial E_{hid}(n)}{\partial \mathbf{w}_j} \right]^{-1} \right| \tag{2.120}$$

The derivation of the upper bound of λ is briefly expressed in the Appendix 2.1.

Besides, the adjustment of the penalty factor λ is employed to control the global search ability and is adapted according to the following conditions:

1. When the convergence is stuck in local minimum, the penalised optimisation is introduced. λ(n) decreases gradually by 0.3% of the λ(n-1) to provide an uphill force to escape from the local minimum.
2. After running the penalised optimisation for few iterations, the λ(n) should be increased by 1% of the λ(n-1) to diminish the effect of the penalty term when the training error starts to decrease.

The following are shown some illustrations by making use of this algorithm in different benchmark problems. First, Fig. 2.4a shows a 3-D error surface plot of modified XOR problem. This modified XOR problem was used to illustrate the problem of local minima during training with the BP algorithm. It is a classical XOR problem but with one more introduced pattern such that a unique global minimum exists. As there is one known local minimum in the error surface, the plot of searching trajectory along the error surface. The network architecture of 2-2-1 with only 9 connection weights was used for this problem. In the range shown, the problem has three minima, one of which is the global minimum. Using a search range of [-5, 5], Fig. 2.4 illustrates the search trajectories convergence. The 2-D contour plot and the comparative performances by the Gaussian penalty function, the Rayleigh penalty function and the Laplace penalty function are also shown in Figs. 2.4. Clearly, similar performance was obtained and the trajectories are able to pull out from the local minimum with the first hundred iterations. The subsequent trajectories are able to follow the surface contour down to the global minimum. The second problem used by this algorithm was the two-spiral benchmark test. This test is reputable to be an extremely and demanding classification problem for any algorithm in the network training. The cascade-correlation could be used to solve this problem using a structure of 2-5-5-5-1 network (three hidden layers with five units of each, "shortcuts" connections required). The network architecture is of high-order structure. Their results are promising but extremely long training time was required despite working under a powerful workstation machine. Fig. 2.5 shows that the solutions yielded by this heuristic hybrid algorithm using different types of penalty terms.

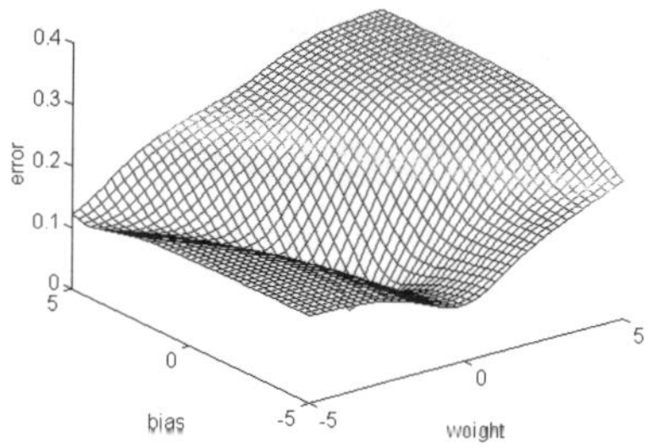

Error surface plot by modified XOR

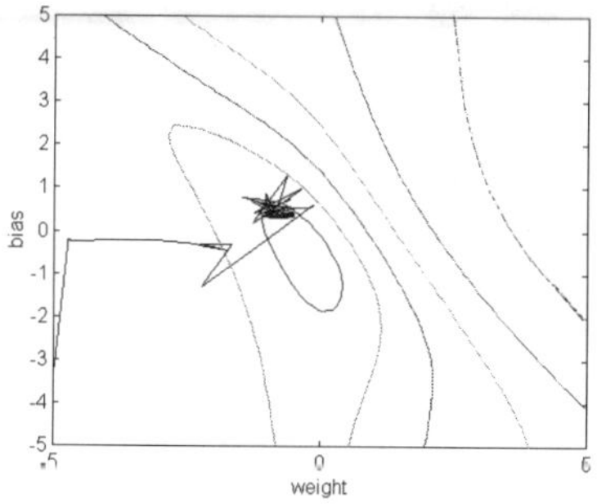

Gaussian penalty approach

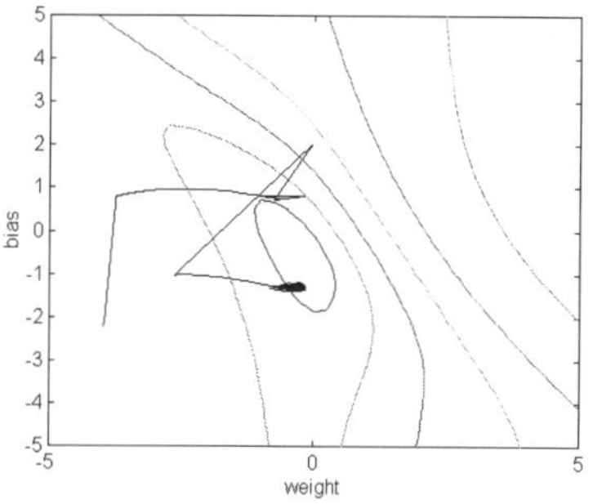

Rayleigh penalty approach

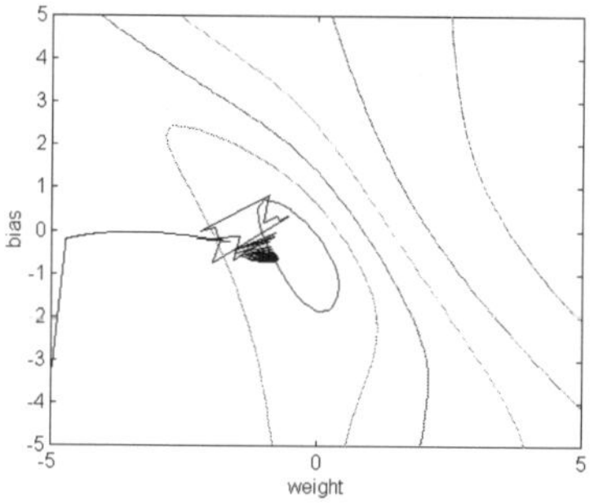

Laplace penalty approach

Figure 2.4. 3-D error plot and 2-D contour plots with searching trajectories of the heuristic global learning algorithm for the modified XOR problems. In these figures shown, the convergence is stuck in the local minimum and the trajectories can be pulled out from the local minimum by the penalty force to converge to the global minimum eventually

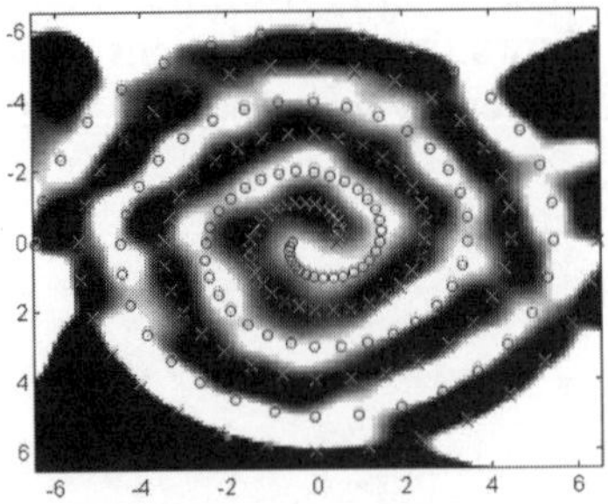

Gassian penalty approach

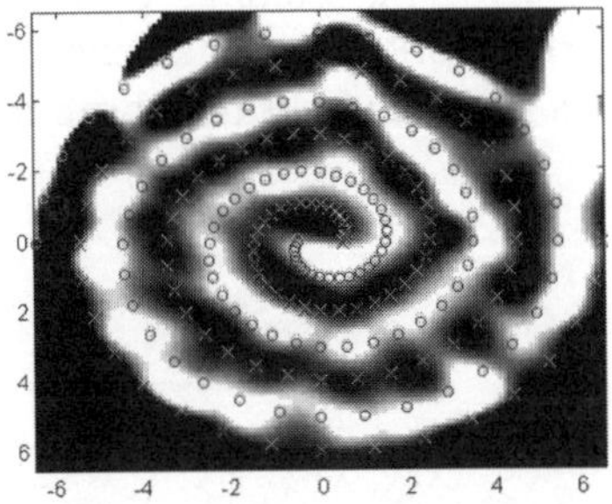

Laplace penalty approach

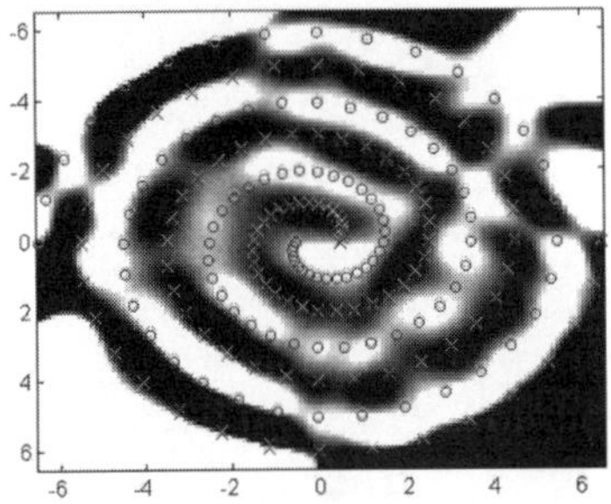

Rayleigh penalty approach

Figure 2.5. Output classification contour images for two-spiral problem by the heuristic global algorithm

2.6. Concluding Remarks

The classified problems affecting the learning performance and the enhancement were addressed in this chapter. Since the learning algorithm is based on the gradient descent typed characteristics (i.e. Backpropagation algorithm), two common problems, i.e. slow convergence speed and susceptibility of local minima, do often occur amid the learning process. This chapter suggests several methods to enhance the learning performance in the neural networks

2.6.1. Fast Learning Algorithms

Three fast learning algorithms have been described to reduce the training time of feedforward neural networks. These three fast training algorithms are Extended Backpropagation (EBP) algorithm, Modified Least Squares Based (MLSB) algorithm, and Extended Least Squares Based (ELSB) Algorithm. The EBP algorithm is based on the adaptation of the learning rate and the momentum coefficient according to the coefficient of correlation between the downhill gradient and the previous weight update. The EBP algorithm greatly reduces the training time when it is compared to the Adaptive Backpropagation algorithm. However, this algorithm is more suitable for the problems with binary target values because the abrupt change in learning rate usually drives the neurons to their extreme values where the binary targets lie. To further increase the rate of convergence, the MLSB algorithm, which is based purely on a linear algebraic method, was described. Although the MLSB algorithm trains the networks to reach small error levels in the first few iterations, the network error cannot be further reduced afterwards. The problem arises from the use of a transformation matrix to transform the optimal output of the hidden layer into the range of the activation function. Apparently, there are numerous ways to construct a transformation matrix in accordance with different distribution of training data. For the problems that require the networks with very high accuracy, the ELSB algorithm was developed. The ELSB algorithm combines the methods used in the EBP algorithm and the MLSB algorithm.

2.6.2. Weight Initialization Methods

Determining the optimal initial weights of feedforward neural networks is another approach to enhance the learning performance. Three weight initialization algorithms, which all are based purely on linear algebraic methods, have been described. In the first weight initialization method (WIA I), the system is first assumed to be linear. This enables the parameters relating the input and output to be evaluated using a least squares method. The initial weights in each layer are then evaluated from these parameters by QR factorization. The algorithm greatly reduced the initial network errors, but this algorithm cannot be applied to the neural networks in which the number of hidden neurons is smaller than the number of input neurons plus one. Another weight initialization algorithm (WIA II), was thus developed to overcome this limitation. In the WIA II algorithm, the outputs of hidden layers are assigned with values in the range in which the derivative of activation function has a large value. The optimal weights are evaluated by the least squares method. Although the WIA II algorithm can be applied to any network architecture, the networks initialized with this algorithm had greater chances of getting stuck when they are further trained. The third weight initialization algorithm (WIA III) was given to improve the second algorithm. The WIA III algorithm determines the distribution of initial values of the weights of all layers except the output layer; the weights connecting to the output layer are evaluated by the least squares algorithm. The WIA III overcomes the shortcomings of the first two weight initialization algorithms and was combined with the ELSB algorithm to form the fast training mechanism. The training mechanism greatly reduces the training time and has additional advantages over other well-known fast training algorithms in terms of the low computational complexity and the storage requirement.

2.6.3. Global Learning Algorithms

Another important issue of the local minima problem was addressed in this chapter. In the past decade, many researchers have proposed different types of algorithms which can provide a significant reduction in the total number of iterations by exploiting the fast convergence characteristics, but getting stuck in the local minima is inevitable in many cases. Convergence stalling is yielded especially in applying to the complicated problems. Random Search (RS) and Simulated Annealing (SA) methods are one of the stochastic typed global optimization methods. In terms of their working mechanism, the approaches are reliable and simply to implement, but larger neural network size is required for solving very complicated classification problems. Deterministic algorithms, such as NOVEL and reactive tabu search, have some advantages over the stochastic versions. They find all deep minima contained in some bound region of the parameter space. Unfortunately finding a solution for 4 hidden units required a total of one week of workstation time. Therefore, a hybrid learning scheme is one of the possible solution to tackle the above problems. This algorithm converges to the global minimum in the probabilistic sense. The methodology of the learning scheme is based on a hybrid of the least squares method and the well-known stochastic optimization method. The least square method determines the weights connected between the output layer and the hidden layer. As a result, the convergence speed is spectacular because the weights are determined in a single step. The stochastic part was employed by the penalized optimization. The idea of this penalty approach is defined by superimposing a discontinuous type function under the weight space domain while the learning process is getting stuck. These studies included the use of Gaussian, Rayleigh, and Laplace functions as the discontinuous penalty functions.

Appendix 2.1.

Assume the weight change $\Delta \mathbf{w}_j(n)$ is the weight difference between $\mathbf{w}_j(n)$ and $\mathbf{w}_j(n-1)$, so

$$\Delta \mathbf{w}_j(n) = \eta\left(-\frac{\partial E_{hid}(n)}{\partial \mathbf{w}_j} + \frac{\partial E_{pen}(n)}{\partial \mathbf{w}_j}\right)$$

$$= \eta\begin{pmatrix} -\dfrac{\partial E_{hid}(n)}{\partial \mathbf{w}_j} + \dfrac{2}{\lambda}\left|\mathbf{w}_j(n-1) - \mathbf{w}_j^*\right| \\ \exp\left(-\dfrac{\left(\mathbf{w}_j(n-1) - \mathbf{w}_j^*\right)^T\left(\mathbf{w}_j(n-1) - \mathbf{w}_j^*\right)}{\lambda^2}\right)\end{pmatrix} \tag{A2.1}$$

Therefore, from (2.118),

$$\Delta E_{gbl}(n) = E_{gbl}(n) + \eta\left[\frac{\partial E_{gbl}(n)}{\partial \mathbf{w}_j}\right]^T\left(-\frac{\partial E_{hid}(n)}{\partial \mathbf{w}_j} + \frac{2}{\lambda}\left|\mathbf{w}_j(n-1) - \mathbf{w}_j^*\right|\right.$$
$$\left.\exp\left(-\frac{\left(\mathbf{w}_j(n-1) - \mathbf{w}_j^*\right)^T\left(\mathbf{w}_j(n-1) - \mathbf{w}_j^*\right)}{\lambda^2}\right)\right) - E_{gbl}(n)$$

$$= \eta\left\|\begin{matrix} -\dfrac{\partial E_{hid}(n)}{\partial \mathbf{w}_j} + \dfrac{2}{\lambda}\left|\mathbf{w}_j(n-1) - \mathbf{w}_j^*\right| \\ \exp\left(-\dfrac{\left(\mathbf{w}_j(n-1) - \mathbf{w}_j^*\right)^T\left(\mathbf{w}_j(n-1) - \mathbf{w}_j^*\right)}{\lambda^2}\right)\end{matrix}\right\|^2 \tag{A2.2}$$

where $\|\cdot\|$ is a norm of the vector.

By the Convergence Theorem from Eq. (2.118), $\Delta E_{gbl}(n) \le 0$ and assume η is positive constant, so,

$$-\frac{\partial E_{hid}(n)}{\partial \mathbf{w}_j} + \frac{2}{\lambda}\left|\mathbf{w}_j(n-1) - \mathbf{w}_j^*\right| \exp\left(-\frac{\left(\mathbf{w}_j(n-1) - \mathbf{w}_j^*\right)^T \left(\mathbf{w}_j(n-1) - \mathbf{w}_j^*\right)}{\lambda^2}\right) \le 0 \tag{A2.3}$$

becomes,

$$\exp\left(-\frac{\left(\mathbf{w}_j(n-1) - \mathbf{w}_j^*\right)^T \left(\mathbf{w}_j(n-1) - \mathbf{w}_j^*\right)}{\lambda^2}\right) \le \frac{\partial E_{hid}(n)}{\partial \mathbf{w}_j}\left(\frac{2}{\lambda}\left|\mathbf{w}_j(n-1) - \mathbf{w}_j^*\right|\right)^{-1}$$

$$-\frac{\left(\mathbf{w}_j(n-1) - \mathbf{w}_j^*\right)^T \left(\mathbf{w}_j(n-1) - \mathbf{w}_j^*\right)}{\lambda^2} \le \log_e\left(\frac{\partial E_{hid}(n)}{\partial \mathbf{w}_j}\left(\frac{2}{\lambda}\left|\mathbf{w}_j(n-1) - \mathbf{w}_j^*\right|\right)^{-1}\right)$$

$$\vdots$$

$$-\left(\begin{array}{l}\dfrac{\left(\mathbf{w}_j(n-1) - \mathbf{w}_j^*\right)^T \left(\mathbf{w}_j(n-1) - \mathbf{w}_j^*\right)}{\lambda^2} \\ + \log_e\left(\dfrac{2}{\lambda}\left|\mathbf{w}_j(n-1) - \mathbf{w}_j^*\right|\right)\end{array}\right) \ge -\log_e\left(\frac{\partial E_{hid}(n)}{\partial \mathbf{w}_j}\right) \tag{A2.4}$$

initially, we assume that $|\lambda| > 0$ and is a quite large number, so $\lambda^2 \to \infty$ then $\frac{1}{\lambda^2} \to 0$,

$$\frac{2}{\lambda_{UB}}\left|\mathbf{w}_j(n-1) - \mathbf{w}_j^*\right| \ge \left[\frac{\partial E_{hid}(n)}{\partial \mathbf{w}_j}\right]$$

$$\therefore \quad \lambda_{UB} \le \left| 2\left|\mathbf{w}_j(n-1) - \mathbf{w}_j^*\right| \left[\frac{\partial E_{hid}(n)}{\partial \mathbf{w}_j}\right]^{-1} \right| \tag{A2.5}$$

then Eq. (2.120) follows.

Exercises

Q2.1. Consider a simple multilayer perceptron (MLP) as shown in Fig. 2.6, compute the individual steps of the back-propagation algorithm. Consider the learning rate to be η=0.5 and sigmoid activation functions $\sigma(h)=\frac{1}{1+e^{-2\beta h}}$ in the nodes with gain β=0.5. Calculate the complete cycle with the input pattern (0, 1) and the input pattern (1, 0).

w0	w1	w2	w3	w4	x1	x2	Oh	Oo	Δw4	Δw3	Δw2	Δw1	Δw0
0.1	-0.1	0	0.1	0.1	0	1							
					1	0	-	-	-	-	-	-	-

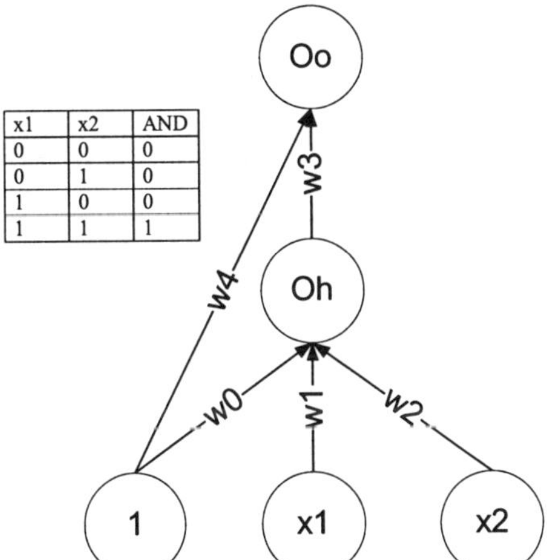

Figure 2.6. MLP network for the AND function

Q2.2. A neural network is being trained on the data for XOR problem. The architecture and the values of the weights and biases are shown in Fig. 2.7.

Using the sigmoid function $\sigma(h) = \frac{1}{1+e^{-2\beta h}}$ with gain β=0.5, compute the activations for each of the units when the input vector (0, 1) is presented. Find the delta factors for the output and hidden units. Using a learning rate of η=0.1, compute the weight corrections. Find the new weights (and biases).

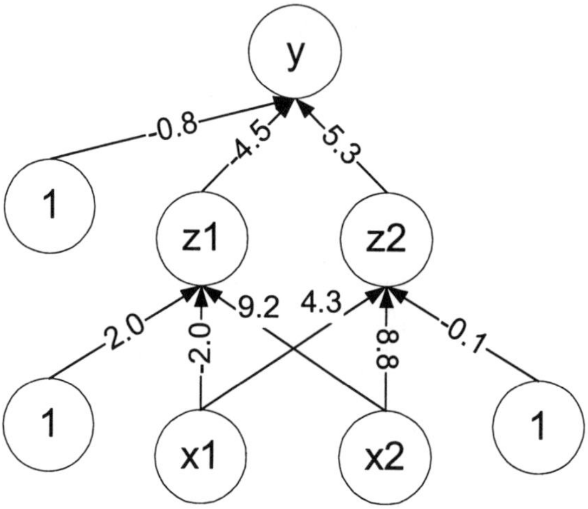

Figure 2.7. Neural network for Q2.2

Q2.3. Repeat the question of Q2.2 for the input vector (1, 0).

Q2.4. Interpret the differences between the weight changes on the connection to the output unit and the weight changes to the hidden units in Q2.2 and Q2.3.

Q2.5. Consider an MLP with three inputs, one output and one hidden layer with 10 nodes. As learning algorithm uses backpropagation and sigmoid functions in the nodes, compute the initial weights at the hidden and output layers using the Weight Initialization Algorithm I as described in Section 2.4.1.

Input patterns			Output
x1	x2	x3	y
0	0	0	0
0	0	1	1
0	1	0	1
0	1	1	0
1	0	0	1
1	0	1	0
1	1	0	0
1	1	1	1

Q2.6. Determine the upper bound of the penalty factor λ by means of convergence proof as similar to equation (2.121) if a Rayleigh penalty term is used for penalized optimization.

Q2.7. Repeat the question of Q2.6 of a Laplace penalty term is used for penalized optimization.

Chapter 3

Generalization and Performance Enhancement

This chapter focuses on the issue of generalization capability of neural network. In this chapter, neural network and network are considered as interchangeable terms. The neural networks possess two paramount features of the universal function approximating (Hornik et al., 1989) and generalization capabilities (Bishop, 1995), (Leung, 1998). Therefore, researchers and engineers often consider neural networks as "black box" type of tools. However, the network performance often varies greatly depending on many factors such as learning algorithms, network size....etc. The network performance is often determined according to a training error and a test error. The training error is a quantity to indicate the closeness between the neural network and the training patterns. The test error is an index to reflect the distance between the neural network and the underlying function based on the unseen test set.

As neural networks are considered universal approximators (Funahashi, 1989), the network training is, in fact, a process of non-parametric functional estimation in the statistical sense. The network training only based on a finite number of training examples is basically an ill-posed problem (Geman et al., 1992). Due to the limited availability of the training samples, there are enormous possible realizations of neural networks delivering comparable level of training error. However, most of the realizations often do not generalize the training data as expected. Only a few realizations are able to generalize the training data as what the underlying function is perceived.

There is no doubt that obtaining an appropriate realization from the enormous possibilities is a baffling problem. Such behavior is analogous to the situation of curve fitting, where the right complexity is needed for a good fit to both training and test sets, as illustrated in Fig. 3.1. The main concern of this chapter is that the error level to the novice examples should be comparable to the training data in a particular application. The trained networks should have a high degree of generalization capability, thus the training error should be a relatively reliable measure of the network performance when the training data is assumed to be sufficient in representing the underlying function.

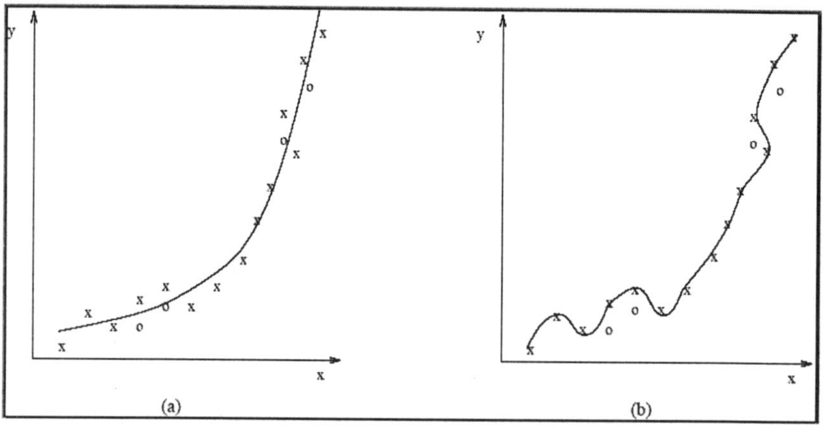

Figure 3.1. (a) A good fit; (b) Over-fit data: Perfect fir on training set, but poor fit on test patterns (x's represent training patterns, o's represents test patterns)

Recently, there are many techniques proposed to enhance the generalization capability. These techniques can be roughly divided into two categories, structure-based and application-based. The structure-based techniques are currently the mainstream approach, which is based on finding an appropriate network structure. The structure-based techniques consist of pruning methods and construction methods. Pruning methods start from initial architecture that is over-parameterized, and selectively removing units and/or connections during the network training, until a satisfactory performance level is reached. Optimal brain damage (LeCun et al., 1990), optimal brain surgeon (Hassibi et al., 1993), weight decay (Krogh & Hertz, 1992), and weight elimination (Weigend et al., 1991) are some of the well-known pruning methods. Construction methods, namely dynamic node creation (Ash,

1989), and Cascade-correlation (Fahlman & Lebiere, 1990), concentrate on incrementally building a network from a simple structure.

It is believed that the choice of the network realizations should also depends upon the nature of the applications. For example, the pattern recognition problems and the regression problems introduce different aspects of requirements. The inclusion of a priori knowledge of a particular application facilitates the network training to converge to a desirable neural network realization. The application-based techniques concentrate on including the a priori knowledge in the network training. This chapter focuses on discussing this issue.

This chapter is organized as follows: Section 3.1 presents some basic properties of cost function as well as generalization capability for neural network optimization. Under simplifying assumptions, the Least-Square (LS) cost function is the most popular used for the feed-forward neural network. However, the LS based optimization algorithms suffer from a problem of "overfitting" noise. Therefore, Section 3.2 focuses on higher order cumulants based cost function, which is blind to Gaussian noise to tackle the problems of the LS cost function. An interesting property of higher order cumulants applying to neural network learning is described in this section. Following that, Section 3.3 presents the concepts of regularization methods based upon higher-order cumulants based cost functions. The methods of adaptive regularization parameters selection are described and suggested how the regularization parameter selects to avoid trapping into sub-optimal solution in this section. An evaluation by synthetic function mapping is also presented in this section. Finally, Section 3.4 presents the concluding remarks of this chapter.

3.1. Cost Function and Performance Surface

In principle, the primary goal of the network training is to model the underlying function of the training data, so that the best possible estimation for the network outputs can be made when the trained neural network is presented with a novice value for the network inputs. The optimal weights obtained by the network training are totally based on the information from the selected objective function. Hence, the selection of objective functions is decisive to the efficiency of the network training and the performance of the trained neural network. There is a common view that different applications may emphasize on different aspects. For time-series forecasting problems, the basic goal of the network training is to model the conditional distribution of the output variables, given the input variables. This motivates the use of the Least Squares (LS) error function that achieves extensive use in the neural based time-series forecasting.

3.1.1. Maximum Likelihood Estimation

For regression problems, the major objective in the network training is not to memorize the training data, but rather to model the underlying function of the data. Hence, the best possible forecasts of the output vector **t** can be made when the trained network is subsequently presented with a novice value for the input vector **x**, viz.

$$\hat{\mathbf{t}} = \arg\max_{\mathbf{t}}\{p(\mathbf{x},\mathbf{t})\} = F(\mathbf{x},\mathbf{W}) \tag{3.1}$$

where $\hat{\mathbf{t}}$ is the estimated **t**; **W** is the network weight vector; F is the function of the neural network; $\arg\max_{\mathbf{x}}\{V(x)\}$ is the value of x that maximizes $V(x)$ and $p(\mathbf{x},\mathbf{t})$ is the joint probability density function. We consider that there is a set of training data $\mathbf{D} = \{(\mathbf{x}_k,\mathbf{t}_k)\}$ and each data sample $(\mathbf{x}_k,\mathbf{t}_k)$ is drawn randomly and independently from the same distribution. In accordance with Maximum Likelihood Estimation, the optimal weight vector **W** is found by maximizing a likelihood function, which is given by

$$\begin{aligned}\zeta &= \prod_k p(\mathbf{x}_k,\mathbf{t}_k)\\ &= \prod_k p(\mathbf{t}_k|\mathbf{x}_k)p(\mathbf{x}_k)\end{aligned} \tag{3.2}$$

where $p(\mathbf{t}_k|\mathbf{x}_k)$ is the probability density of **t** given a particular input vector **x**; and $p(\mathbf{x})$ is the unconditional density of **x**. The likelihood function ζ can be converted into an objective function

$$\varepsilon = -\ln\zeta = -\sum_k \ln p(\mathbf{t}_k|\mathbf{x}_k) - \sum_k \ln p(\mathbf{x}_k) \tag{3.3}$$

As the second term is equation 3.3 is independent of the neural network function F, we can redefine the objective function ε by

$$\varepsilon = -\sum_k \ln p(\mathbf{t}_k|\mathbf{x}_k) \tag{3.4}$$

Hence, we can obtain a maximum likelihood estimation based on the equation 3.4.

3.1.2. *The Least-Square Cost Function*

For the sake of analysis, we only consider the neural network with a single linear output because the extension to the multiple outputs is obvious. Assume the case have *n* data points $(\mathbf{x}_k, \mathbf{t}_k)$ where *k*=1,2,…,*n*, and suppose that the data points $(\mathbf{x}_k, \mathbf{t}_k)$ are statistically independent. We consider the target value t_k is given by an unknown function *h* with added Gaussian noise perturbation *e*, viz.

$$t_k = h(\mathbf{x}_k) + e_k \tag{3.5}$$

where e_k is Gaussian distributed with zero mean and standard deviation σ. The noise *e* is also statistically independent of the input vector **x** and **t**. As the neural network $F(\mathbf{x}, \mathbf{W})$ is used to model the unknown function $h(\mathbf{x})$, thc conditional probability density $p(t|x)$ is given by

$$p(t|\mathbf{x}) = \frac{1}{\sqrt{2\pi\sigma^2}} \exp\left(-\frac{\{t - F(\mathbf{x}, \mathbf{W})\}^2}{2\sigma^2} \right) \tag{3.6}$$

According to Maximum Likelihood Method, the error function *E* in the equation 3.5 will be

$$\begin{aligned} \varepsilon &= -\sum_{k=1}^{n} \ln p(t_k|\mathbf{x}_k) \\ &= \frac{1}{2\sigma^2} \sum_{k=1}^{n} \{t_k - F(\mathbf{x}_k, W)\}^2 + n \ln \sigma + \frac{n}{2} \ln(2\pi) \end{aligned} \tag{3.7}$$

As the second and third terms of equation 3.7 are independent of weights **W**, we obtain the familiar expression for the Least-Sqaures (LS) error function

$$\varepsilon = a \sum_{k=1}^{n} \{t_k - F(\mathbf{x}_k, \mathbf{W})\}^2 \tag{3.8}$$

where a is the scalar constant. Now, let us consider the asymptotic condition of the error function when the size n of the training data set goes to infinity. By selecting $a = \frac{1}{n}$, we have

$$\begin{aligned}\varepsilon &= \lim_{n\to\infty}\frac{1}{n}\sum_{k=1}^{n}\{t_k - F(\mathbf{x}_k,\mathbf{W})\}^2 \\ &= \iint\{t_k - F(\mathbf{x}_k,\mathbf{W})\}^2 p(t,\mathbf{x})dtd\mathbf{x} \\ &= E\left\{\{t_k - F(\mathbf{x}_k,\mathbf{W})\}^2\right\}\end{aligned} \tag{3.9}$$

We now substitute $t - F(\mathbf{x},\mathbf{W}) = t - E\{t|\mathbf{x}\} + E\{t|\mathbf{x}\} - F(\mathbf{x},\mathbf{W})$ in the above equations,

$$\begin{aligned}\varepsilon &= E\left\{\{t - E\{t|\mathbf{x}\} + E\{t|\mathbf{x}\} - F(\mathbf{x},\mathbf{W})\}^2\right\} \\ &= E\left\{\{t - E\{t|\mathbf{x}\}\}^2\right\} + E\left\{\{E\{t|\mathbf{x}\} - F(\mathbf{x},\mathbf{W})\}^2\right\} + E\{t - E\{t|\mathbf{x}\}\}E\{E\{t|\mathbf{x}\} - F(\mathbf{x},\mathbf{W})\}\end{aligned} \tag{3.10}$$

Since the third term vanishes, we have

$$\varepsilon = E\left\{\{t - E\{t|\mathbf{x}\}\}^2\right\} + E\left\{\{E\{t|\mathbf{x}\} - F(\mathbf{x},\mathbf{W})\}^2\right\} \tag{3.11}$$

Because the first term of equation 3.11 is independent of the network weights **W**, the absolute minimum of the error function in equation 3.11 when the second term vanishes, which corresponds to the following result

$$F(\mathbf{x},\mathbf{W}^*) = E\{t|\mathbf{x}\} \tag{3.12}$$

where $\mathbf{W}^*$ is the network weights at the absolute minimum of the error function. Based on the derivation of the LS error function, the distribution of the target value t_k is assumed to be Gaussian. A maximum likelihood estimate of the unknown function h can be obtained when the following conditions are met:

1. The data set must be sufficiently large that it approximates an infinite data set;
2. The neural network $F(\mathbf{x},\mathbf{W})$ must be sufficiently general that there exists a choice of parameters which makes second term in equation 3.11 sufficiently small;

The optimization of the network weights is without getting stuck in the local minima of the LS error function.

Conditions 1 and 2 sometimes may not be met together at the same time. The size of the data set required by Condition 1 depends upon the size of the neural network. In order to have a generalized neural network, a larger network size is required, thus the data size has to become larger accordingly. The data sampling in many applications is extremely expensive and difficult to obtain. The training process with a large data set is sometimes impossible due to the enormous computational effort involved. We have to consider the network generalization capability when the data set may no longer be sufficiently large as described as Condition 1.

Now, we consider there is a finite data set $D = \{(\mathbf{x}_1, t_1), \ldots, (\mathbf{x}_n, t_n)\}$ consisting of n patterns used to determine the neural network $F(\mathbf{x}, \mathbf{W})$. Consider the mean-squared error of the neural network $F(\mathbf{x})$ as an estimator of $E\{t|\mathbf{x}\}$ based on the finite data set D. The mean-squared error ε_D is defined by

$$\varepsilon_D = E_D\{F(\mathbf{x}) - E\{t|\mathbf{x}\}\} \tag{3.13}$$

where $E_D\{\cdot\}$ denotes the expectation over all the patterns in the data set D. We now substitute

$$F(\mathbf{x}) - E\{t|\mathbf{x}\} = F(\mathbf{x}) - E_D\{F(\mathbf{x})\} + E_D\{F(\mathbf{x})\} - E\{t|\mathbf{x}\} \tag{3.14}$$

in ε_D and manipulate it as follows:

$$\begin{aligned} \varepsilon_D &= E_D\left\{\{F(\mathbf{x}) - E_D\{F(\mathbf{x})\}\}^2\right\} \\ &\quad + E_D\left\{\{E_D\{F(\mathbf{x})\} - E\{t|\mathbf{x}\}\}^2\right\} \\ &\quad + 2\{E_D\{F(\mathbf{x})\} - E\{t|\mathbf{x}\}\}E_D\{F(\mathbf{x}) - E_D\{F(\mathbf{x})\}\} \end{aligned} \tag{3.15}$$

Hence, the mean-squared error ε_D will be

$$\varepsilon_D = E_D\left\{\{E_D\{F(\mathbf{x})\} - E\{t|\mathbf{x}\}\}^2\right\} + E_D\left\{\{F(\mathbf{x}) - E_D\{F(\mathbf{x})\}\}^2\right\} \tag{3.16}$$

The first term of the above equation represents the bias of the neural network $F(\mathbf{x})$ measured with respect to the function $h(\mathbf{x}) = E\{t|x\}$. The second term represents the variance of the neural network function. Thus, there is a cost of large variance for achieving a small bias by a sufficient general neural network function. In other words, when the network training is based on the LS error function and a finite training data set, the problem of poor generalization capability (overfitting the training samples) is likely to occur. Especially, if the training data is perturbed, the trained neural network may overfit the noisy training samples when the size of the perturbed samples is not sufficiently large to make the cross term $E\{t - E\{t|\mathbf{x}\}\}E\{E\{t|\mathbf{x}\} - F(\mathbf{x})\}$ in equation 3.10 to be zero.

As mentioned before, the noise perturbation or the target values of the training data is assumed to be Gaussian normal distributed. On one hand, the LS error function is only in terms of the second-order moment, which cannot distinguish between the Gaussian distribution and any other distribution having the same mean and variance. The distribution of the residual error, $e_k = t_k - F(\mathbf{x}_k, \mathbf{W})$, of the trained network $F(\mathbf{x}, \mathbf{W})$ may not be Gaussian distributed as assumed. This degrades the generalization capability of the trained neural network. On the other hand, the distribution of real world data is non-Gaussian in general. The network training based on the LS error function may not be able to obtain a maximum likelihood estimate as mentioned in Section 3.1.1.

3.2. Higher-Order Statistic Generalization

In the previous section, many problems, especially the problem of the generalization capability, stem from the assumption of the Gaussian distribution of the training data and the LS error function is not able to distinguish any non-Gaussian process from a Gaussian process with the same values of mean and variance. In the real world, a wide variety of applications, including sonar, digital communications, seismology, and radio astronomy, are arisen by non-Gaussian signals. When traditional Gaussian model based techniques, such as the LS error based algorithms, is applied to non-Gaussian signals, the performance is often not satisfactory. To analyze the non-Gaussian data, the use of a much more sophisticated statistical methodology is needed to supplement the deficient of the traditional Gaussian model based techniques. Recently, higher-order statistics were applied to various real world non-Gaussian (or, possible nonlinear) processes. Higher-order statistics have been applied in many areas such as system identification and parameter estimation (Tugnait, 1990), and noise estimation and detection (Sadler & Giannakis, 1994). The general advantages behind the use of higher-order statistics are threefold: 1) to extract information due to deviations from Gaussianness (normality), 2) to estimate the phase of non-Gaussian parametric signals, and 3) to detect and characterize the nonlinear properties of mechanisms which generate time-series via phase relations of their harmonic components (Nikias & Raghuveer, 1987). Hence, a new type of objective functions is proposed based on the higher-order statistics.

3.2.1. Definitions and Properties of Higher-Order Statistics

Higher-Order Statistic (HOS) provides a unique feature of suppressing gaussian noise processes of unknown spectral characteristics (Nilias & Petropuou, 1993). Cumulants of order $r > 2$ are blind to any kind of a Gaussian process (white or color) whereas correlation is not. As a result, cumulants-based methods are able to boost signal-to-noise ratio when non-Gaussian signals are perturbed by Gaussian measurement noise (Mendel, 1991). The definition and the essential properties of higher-order statistics are summarized as follows:

Given a set of n real random variables $\{x_1, x_2, \ldots, x_n\}$, joint moments of order $r = k_1 + k_2 + \cdots + k_n$ are given by (Papoulis, 1991)

$$\begin{aligned} Mom\{x_1^{k_1}, x_2^{k_2}, \ldots, x_n^{k_n}\} &= E\{x_1^{k_1}, x_2^{k_2}, \ldots, x_n^{k_n}\} \\ &= (-j)^r \left.\frac{\partial^r \Phi(\omega_1, \omega_2, \ldots, \omega_n)}{\partial\omega^{k_1} \partial\omega^{k_2} \cdots \partial\omega^{k_n}}\right|_{\omega_1=\omega_2=\cdots=\omega_n=0} \end{aligned} \tag{3.17}$$

where

$$\Phi(\omega_1, \omega_2, \ldots, \omega_n) = E\{\exp j(\omega_1 x_1 + \cdots + \omega_n x_n)\} \tag{3.18}$$

is their joint characteristic function. The joint cumulants of order r of the same set random variables are defined as

$$Cum\{x_1^{k_1}, x_2^{k_2}, \ldots, x_n^{k_n}\} = (-j)^r \left.\frac{\partial^r \ln \Phi(\omega_1, \omega_2, \ldots, \omega_n)}{\partial\omega^{k_1} \partial\omega^{k_2} \cdots \partial\omega^{k_n}}\right|_{\omega_1=\omega_2=\cdots=\omega_n=0} \tag{3.19}$$

The general relationship between joint moments of $\{x_1, x_2, \ldots, x_n\}$ and joint cumulants $Cum\{x_1, x_2, \ldots, x_n\}$ of order $r = n$ is given by

$$Cum\{x_1, x_2, \ldots, x_n\} = \sum (-1)^{p-1}(p-1)! \cdot E\left\{\prod_{i \in s_1} x_i\right\} E\left\{\prod_{i \in s_2} x_i\right\} \cdots E\left\{\prod_{i \in s_3} x_i\right\} \tag{3.20}$$

where the summation extends over all partitions $(s_1, s_2, \ldots, s_p)$, $p = 1,2,\ldots,n$, of the set of integer $(1,2,\ldots,n)$. Therefore, the joint cumulants can be expressed in terms of the joint moments of a set of random variables, such as

$$Cum\{x_1\} = Mom\{x_1\} = E\{x_1\} \tag{3.21}$$

$$\begin{aligned} Cum\{x_1^2\} &= Mom\{x_1^2\} - Mom\{x_1\}^2 \\ &= E\{x_1^2\} - (E\{x_1\})^2 \end{aligned} \tag{3.22}$$

$$\begin{aligned} Cum\{x_1^3\} &= Mom\{x_1^3\} - 3Mom\{x_1^2\}Mom\{x_1\} + 2(Mom\{x_1\})^3 \\ &= E\{x_1^3\} - 3E\{x_1^2\}E\{x_1\} + 2(E\{x_1\})^3 \end{aligned} \tag{3.23}$$

$$\begin{aligned} Cum\{x_1^4\} &= Mom\{x_1^4\} - 4Mom\{x_1^3\}Mom\{x_1\} - 3(Mom\{x_1^2\})^2 \\ &\quad + 12Mom\{x_1^2\}(Mom\{x_1\})^2 - 6(Mom\{x_1\})^4 \\ &= E\{x_1^4\} - 4E\{x_1^3\}E\{x_1\} - 3(E\{x_1^2\})^2 \\ &\quad + 12E\{x_1^2\}(E\{x_1\})^2 - 6(E\{x_1\})^4 \end{aligned} \tag{3.24}$$

If $\{x_1\}, t = 0, \pm1, \pm2, \cdots$ is a real stationary random process and its moments and cumulants up to order n exists, then $Mom\{x_t, x_{t+\tau_1}, \ldots, x_{t+\tau_{n-1}}\}$ and $Cum\{x_t, x_{t+\tau_1}, \ldots, x_{t+\tau_{n-1}}\}$ will depend only on the time difference $\tau_i = 0, \pm1, \pm2, \cdots$ for all i. The variance, shewness, and kurtosis measures of the distribution of $\{x_t\}$ can be in terms of $Cum\{x_t^2\}$, $Cum\{x_t^3\}$ and $Cum\{x_t^4\}$ respectively. Moreover, a process $\{x_t\}$ is said to be ergodic in the most general form if the expected value $E\{\cdot\}$ can be computed by time averages, viz.

$$E\{x_t, x_{t+\tau_1}, \ldots, x_{t+\tau_{n-1}}\} = \lim_{M\to\infty} \frac{1}{2M+1} \sum_{t=-M}^{+M} x_t x_{t+\tau_1} \cdots x_{t+\tau_{n-1}} \quad (3.25)$$

3.2.2. The Higher-Order Cumulants based Cost Function

In many real world applications, noise perturbation is inevitable and it may significantly affect the network training when the size of the training data is not sufficiently large. To diminish the undesired noise effect, a higher-order cumulant (HOC) based cost function was proposed (Leung & Chow, 1997) to resolve the problem of poor generalization when the network is working under a noisy condition and the data size N is small compared to the number of weights. This new cost function provides useful features suppressing Gaussian noise processes of any spectral characteristics and boosting the signal-to-noise ratio when non-Gaussian signals are perturbed by Gaussian measurement noises. The fourth-order cumulant (FOC) cost function is defined by

$$H_{FOC} = E_D\{(t - F(\mathbf{x}, \mathbf{W}))^2\} + \lambda\left(Cum\{(t - F(\mathbf{x}, \mathbf{W}))^4\}\right)^2 \quad (3.26)$$

where $E_D\{\cdot\}$ is the expected value over all patterns on the training data set $D = \{(\mathbf{x}_k, t_k)\}, k = 1,2,\cdots,n; Cum_D\{\cdot\}$ is the value of the cumulants over all patterns on D; and λ is the positive regularization parameter.

Let $F(\mathrm{x}, \mathrm{W})$ be a sufficient general feedforward neural network with a single output node. Suppose an unknown function $h(\mathbf{x})$ of the target system governs the mappings from the input vector $\mathbf{x}$ of dimension m to the output value $y = h(\mathbf{x})$ such that

$$t = y + e \quad (3.27)$$

where e is the additive noise perturbation with zero mean and finite variance. A finite number of samples from the target system are collected for the network training. We assume that the data sample $(\mathbf{x}_k, t_k)$ is independent and identically distributed, and the noise perturbation e is independent of the system output y. Now, we consider the asymptotic condition of the FOC objective function. When the size of the training set approaches infinity, by substituting

$$t - F(\mathbf{x}, \mathbf{W}) = t - E\{t|\mathbf{x}\} + E\{t|\mathbf{x}\} - F(\mathbf{x}, \mathbf{W})$$

we have the FOC objective function

$$
\begin{aligned}
H_{FOC} &= E\left\{(t - F(\mathbf{x}, \mathbf{W}))^2\right\} + \lambda\left(Cum\left\{(t - F(\mathbf{x}, \mathbf{W}))^4\right\}\right)^2 \\
&= E\left\{\left((t - E\{t|\mathbf{x}\}) + (E\{t|\mathbf{x}\} - F(\mathbf{x}, \mathbf{W}))\right)^2\right\} \\
&\qquad + \lambda\left(Cum\left\{\left((t - E\{t|\mathbf{x}\}) + (E\{t|\mathbf{x}\} - F(\mathbf{x}, \mathbf{W}))\right)^4\right\}\right)^2 \\
&= E\left\{(E\{t|\mathbf{x}\} - F(\mathbf{x}, \mathbf{W}))^2\right\} + E\left\{(t - E\{t|\mathbf{x}\})^2\right\} \\
&\qquad + \lambda\left(Cum\left\{(E\{t|\mathbf{x}\} - F(\mathbf{x}, \mathbf{W}))^4\right\} + Cum\left\{(t - E\{t|\mathbf{x}\})^4\right\}\right)^2
\end{aligned}
\tag{3.28}
$$

Because the terms $E\left\{(t - E\{t|\mathbf{x}\})^2\right\}$ and $Cum\left\{(t - E\{t|\mathbf{x}\})^4\right\}$ are independent of the network weights **W**, the absolute minimum of the FOC objective function in equation 3.28 at $\mathbf{W}^*$ when the terms $E\left\{(E\{t|\mathbf{x}\} - F(\mathbf{x}, \mathbf{W}))^2\right\}$ and $Cum\left\{(E\{t|\mathbf{x}\} - F(\mathbf{x}, \mathbf{W}))^4\right\}$ vanish. This corresponds to the following result

$$
F(\mathbf{x}, \mathbf{W}^*) = E\{t|\mathbf{x}\} \tag{3.29}
$$

as the case of the LS error function. Thus, when the output variable t or the noise perturbation e are Gaussian distributed, a maximum likelihood estimate $F(\mathbf{x}, \mathbf{W}^*)$ can be obtained by applying the FOC objective function. Also, the signal-to-noise ratio is boosted because the term $Cum\left\{(t - E\{t|\mathbf{x}\})^4\right\}$ will be zero if the noise perturbation is Gaussian distributed. It is believed that the inclusion of the fourth-order cumulant term is capable of facilitating the network training because the term can extract more high-order information from the training data. The fourth-order cumulant term also enables the network training to capture the phase information from the data, which is essential to time-series forecasting.

The foremost situation is when the training data set D is finite. Now, we consider there is a finite data set D consisting of n patterns which is used to determine the neural network $F(\mathrm{x}, \mathrm{W})$. Consider the fourth-order cumulant of the error of the neural network $F(\mathbf{x})$ as an estimator of $E\{t|\mathbf{x}\}$ based on the finite data set D. The fourth-order cumulant of the error is defined by

$$\varepsilon_{D,FOC} = Cum_D\left\{(F(\mathbf{x}) - E\{t|\mathbf{x}\})^4\right\} \tag{3.30}$$

We now substitute equation 3.14 in $\varepsilon_{D,FOC}$ and we have

$$\begin{aligned}\varepsilon_{D,FOC} &= Cum_D\left\{\left((F(\mathbf{x}) - E_D\{F(\mathbf{x})\}) + (E_D\{F(\mathbf{x})\} - E\{t|\mathbf{x}\})\right)^4\right\} \\ &= Cum_D\left\{(F(\mathbf{x}) - E_D\{F(\mathbf{x})\})^4\right\} + Cum_D\left\{(E_D\{F(\mathbf{x})\} - E\{t|\mathbf{x}\})^4\right\}\end{aligned} \tag{3.31}$$

because $(F(\mathbf{x}) - E_D\{F(\mathbf{x})\})$ and $(E_D\{F(\mathbf{x})\} - E\{t|\mathbf{x}\})$ are independent of each other. Also, neural networks with linear output nodes are commonly applied in regression problems, such as time-series forecasting. We can consider the output of the neural network $\hat{y} = F(\mathbf{x})$ as a random variable expressed in the form of

$$\hat{y} = \sum_{i=1}^{l} v_i \tag{3.32}$$

where $\{v_i\}$, $i = 1,2,\ldots,l$, is the set of the random variables of the inputs of the output node of the neural network. According to the Central Limit Theorem (Papoulis, 1991), the distribution of $\hat{y}$ approaches a Gaussian distribution as l increases. Consequently, it is reasonably assumed that $(F(\mathbf{x}) - E_D\{F(\mathbf{x})\})$ is Gaussian distributed so that the term $Cum_D\left\{(F(\mathbf{x}) - E_D\{F(\mathbf{x})\})^4\right\}$ vanishes. Hence, we have

$$\varepsilon_{D,FOC} = Cum_D\left\{(E_D\{F(\mathbf{x})\} - E\{t|\mathbf{x}\})^4\right\} \tag{3.33}$$

In other words, there is no bias/variance dilemma to the cumulant term in the FOC objective function when the size of the neural network is sufficiently large.

As mentioned in the Section 3.1.2, the LS error function or the sum-of-squares error function is only in terms of the second-order moment which can characterize all information of a Gaussian distribution. Second-order moments cannot, nevertheless, distinguish between the Gaussian distribution and any other distribution having the same mean and variance. If the network training is based on the sum-of-squares error function, it is probable to have a non-Gaussian distributed residual error, $e_k = t_k - F(\mathbf{x}, \mathbf{W})$, especially when the size of the training data is not sufficiently large. The trained network may probably overfit the training data. In contrast, the FOC objective function is capable of not only characterizing a non-Gaussian distribution but also measuring the kurtosis or "Gaussianity" of the distribution. Consequently, the FOC objective function enables the network training to filter out the undesired functional estimates of which the residual error is non-Gaussian distributed. The residual error is squeezed to be Gaussian distributed as much as possible. Thus, the trained network exhibits a higher generalization capability when the output of the target system or the noise perturbation is Gaussian distributed.

Furthermore, because the distribution of the residual error of the trained network is very close to Gaussian, the reliability of the network training can be coarsely estimated based on the test error of the novice data set. The confidence upper bound of the approximation error of the trained network,

$$\varepsilon_a = E\left\{(t_k - F(\mathbf{x}, \mathbf{W}))^2\right\} \tag{3.34}$$

can be computed by

$$\varepsilon_a \le \frac{n\varepsilon_t}{\chi^2_{n-1}(\beta)} \tag{3.35}$$

with a confidence level of 100β%, where n is the number of patterns in the test data set $D_t = \{(t_i, \mathbf{x}_i)\}$, $i = 1,2,\ldots,n$; the test error ε_t is given by

$$\varepsilon_t = \frac{1}{n}\sum_{i=1}^{N}(t_i - F(\mathbf{x}_i, \mathbf{W}))^2 \tag{3.36}$$

and the χ^2 distribution is defined as

$$\chi^2_n(x) = \int_x^{\infty} \frac{1}{2^{n/2}\Gamma(n/2)} y^{\frac{n}{2}-1} \exp(-y/2)\,dy \tag{3.37}$$

The detailed deviation of the confidence upper bound is summarized in appendix 3.1.

3.2.3. *Property of the Higher-Order Cumulant Cost Function*

Although many global learning algorithms have proposed to find an optimal solution on an error surface containing sub-optimal solutions, the network training often requires the training duration in terms of days or weeks under the platform of workstation level. The required training time is often prohibitively long for real-world applications. The reason behind the long training time is not difficult to understand. These global learning algorithms are developed for general applications and objective functions. The characteristics of particular objective function and application which may facilitate the network training are not fully utilized. It is believed that utilizing the characteristics enables the network training to speed up and to converge to an optimal solution, or a solution with an acceptable generalization error. In this section, an interesting property between $E\{\varepsilon^2\}$ and $Cum\{\varepsilon^r\}$, $r=3,4$, is found when the neural networks are applied to time-series forecasting. Based on the property, we are able to reconstruct the HOC objective functions such that there is no sub-optimal solution. Consequently, the HOC objective functions together with specified regularization method[1] enable the network training not to be trapped into sub-optimal solutions. The detailed description of the property is shown beneath.

Suppose that a set of training pairs $D=\{(\mathbf{x}_k,t_k)\}$ is available. The stationary time-series $\{t_k\}$ is represented by

$$t_k = g(\mathbf{x}_k)+\eta_k, \qquad \text{for all time step } k \le 0 \tag{3.38}$$

where $g(\mathbf{x}_k)$ is the underlying regression function; η_k is the noise perturbation, which is zero mean and finite *r*-order cumulants, $r=2,3,4$, and η_k is independent and identically distributed. We assume that the underlying function $g(\mathbf{x})$ belongs to the function class $\{F(\mathbf{x},\mathbf{W})\}$. In other words, the selected architecture of the neural network is sufficient to exactly mimic the unknown regression function $g(\mathbf{x}_k)$. Hence, the minimum mean square error optimal predictor of t_k given the value of x_k is the conditional expectation

[1] This regularization method is called Adaptive Regularization Parameter Selection (ARPS), which discusses in Section 3.3.

$$\hat{t}_k = E\{t|\mathbf{x}_k\} = g(\mathbf{x}_k) \tag{3.39}$$

According to the Ergodicity Theorem, the moments of the stationary time-series $\{t_k\}$ can be approximated by the corresponding time average. For this reason, we suppose that without the loss of generality there exists an optimal solution $\mathbf{W}^*$ on the LS error surface such that, for all $\mathbf{W}$,

$$E_k\left\{\left(t_k - F(\mathbf{x}_k, \mathbf{W}^*)\right)^2\right\} \le E_k\left\{\left(t_k - F(\mathbf{x}_k, \mathbf{W})\right)^2\right\} \tag{3.40}$$

where $E_k\{\cdot\}$ is the time average or the expectation over the time step *k*. Therefore, the underlying function $g(\mathbf{x}_k)$ can be exactly modeled by the best-fit neural network $F(\mathbf{x}_k, \mathbf{W}^*)$ as follows: for all time step *k*,

$$t_k = F(\mathbf{x}_k, \mathbf{W}^*) + e_k^* \tag{3.41}$$

where the residual e_k^* equals the noise perturbation η_k, i.e. $e_k^* = \eta_k$. Then, we have, for all weight vector $\mathbf{W}$ and time step *k*,

$$t_k = F(\mathbf{x}_k, \mathbf{W}) + e_k \tag{3.42}$$

and

$$F(\mathbf{x}_k, \mathbf{W}^*) = F(\mathbf{x}_k, \mathbf{W}) + e_{\mathbf{w}k} \tag{3.43}$$

where the residual e_k is given by

$$e_k = e_{\mathbf{w}k} + e_k^* \tag{3.44}$$

the following theorem has been proved.

Theorem 3.1 If there exists a $\mathbf{W}$ such that, $\nabla E_k\{e_k^2\} = 0$ and $\nabla Cum_k\{e_k^r\} = 0$, for r=3,4 then

$$E_k\left\{\left(F(\mathbf{x}_k, \mathbf{W}^*) - F(\mathbf{x}_k, \mathbf{W})\right)^2\right\} = E_k\{e^2{}_{\mathbf{w}k}\} = 0$$

The proof is based on the following assumptions:

[H1] For all w in $\mathbf{W}$, e_k^* and $\dfrac{\partial F(\mathbf{x}_k, \mathbf{W})}{\partial w}$ are statistically independent.

[H2] e_k^* and $e_{\mathbf{w}k}$ are also statistically independent.

[H3] $g(\mathbf{x})$ can be exactly mimicked by the neural network $F(\mathbf{x}, \mathbf{W}^*)$.

[H4] The series t_k is stationary.

The detail of the proof is summarized in Appendix 3.2. The above assumptions H1 and H2 are selected based on the rationale that the error, $g(\mathbf{x}_k) - F(\mathbf{x}_k, \mathbf{W}^*)$, is negligible compared with the noise perturbation η_k in order to have a good network generalization. For the simplicity of analysis, e_k^* is considered to be a noise which is independent and identically distributed with zero mean and finite variance.

According to **Theorem 3.1**, the new objective function (i.e. FOC function) is redefined as

$$H'_{FOC} = E_D\left\{(t - F(\mathbf{x}, \mathbf{W}))^2\right\} + \lambda\left|Cum_D\left\{(t - F(\mathbf{x}, \mathbf{W}))^4\right\}\right| \quad (3.45)$$

In the equation 3.45, the regularized objective function is composed of two terms which are $Cum_k\{e_k^r\}$ and $E_k\{e_k^2\}$. Hence, the network training will stall only when

1. The gradients of $E_k\{e_k^2\}$ and $Cum_k\{e_k^r\}$, $r=3,4$ both vanish, or
2. The sum of the two gradients is zero, i.e. for $r=3,4$

$$\nabla E_k\left\{e_k^2\right\} + \lambda\nabla\left|Cum_k\left\{e_k^r\right\}\right| = 0 \quad (3.46)$$

when the gradients of $E_k\{e_k^2\}$ and $Cum_k\{e_k^r\}$ are nonzero.

Condition 1 is one of the situations that the algorithm finds a minimum. In accordance with **Theorem 3.1**, when Condition 1 occurs, an optimal solution is found. Moreover, Condition 2 is the condition of sub-optimal solutions when the Hessian matrix is semi-positive. As seen in equation 3.46, there exists a parameter λ affecting the condition and the location of sub-optimal solutions on the error surface of the objective functions, while the locations of the optimal solutions are not affected by the magnitude of λ. In other words, when a gradient type training process gets stuck in sub-optimal solutions on the composite surface, we can easily relocate the obstruction (sub-optimal solution) by changing the value of λ. Moreover, theorem 3.1 implies that the locations of the optimal solutions in Condition 1 do not change with λ. The neural network training will not stall at a sub-optimal solution if λ is properly controlled by a regularization selection method, which is discussed in Section 3.4. It is worth to note that more than one optimal solution may exist. Different initial guesses of network weights lead to different downhill trajectories and may end up to different optimal solutions. In practical situations, the network training can converge to the best possible solution under finite available training data and training time. It is believed that the best possible solution may be close to one of the optimal solutions.

3.2.4. Learning and Generalization Performance

A batch-mode HOC objective function (shown in equation 3.26) based backpropagation algorithm is derived in accordance with the fact that the time-series is assumed to be stationary. In fact, when the time-series is non-stationary, this assumption can easily be met by the proper selection of time-series model. Hence, by assuming the time-series to be ergodic, the moments can be approximated by

$$E_D\{e\} \approx \frac{1}{n}\sum_{i=1}^{n}\left(t_{t+i} - F(\mathbf{x}_{t+i}, \mathbf{W})\right) \tag{3.47}$$

$$E_D\{e^2\} \approx \frac{1}{n}\sum_{i=1}^{n}\left(t_{t+i} - F(\mathbf{x}_{t+i}, \mathbf{W})\right)^2 \tag{3.48}$$

$$E_D\{e^3\} \approx \frac{1}{n}\sum_{i=1}^{n}\left(t_{t+i} - F(\mathbf{x}_{t+i}, \mathbf{W})\right)^3 \tag{3.49}$$

$$E_D\{e^4\} \approx \frac{1}{n}\sum_{i=1}^{n}\left(t_{t+i} - F(\mathbf{x}_{t+i}, \mathbf{W})\right)^4 \tag{3.50}$$

where the training data D is $\{(\mathbf{x}_{t+i}, t_{t+i})\}$, $i = 1,2,\cdots,n$. According to the gradient descent optimization technique, the solution $\mathbf{W}$, which minimizes the HOC objective function, can be computed iteratively by the following equation:

$$\begin{aligned}\mathbf{W}(k+1) &= \mathbf{W}(k) + \Delta\mathbf{W}(k) \\ &= \mathbf{W}(k) + \left(\eta\nabla_{\mathbf{W}(k)}H_{FOC} + \beta\Delta\mathbf{W}(k-1)\right)\end{aligned} \tag{3.51}$$

where k is the iteration number; η is the learning factor; β is the momentum factor; **W**(0) is randomly initialized and

$$\begin{aligned}\nabla_{\mathbf{W}(k)} H_{FOC} = {} & \nabla_{\mathbf{W}(k)} E_D\left\{e^2\right\} + \lambda Cum_D\left\{e^4\right\}(\nabla_{\mathbf{W}(k)} E_D\left\{e^4\right\} \\ & -4E_D\left\{e\right\}\nabla_{\mathbf{W}(k)} E_D\left\{e^3\right\} - 4E_D\left\{e^3\right\}\nabla_{\mathbf{W}(k)} E_D\left\{e\right\} \\ & -6E_D\left\{e^2\right\}\nabla_{\mathbf{W}(k)} E_D\left\{e^2\right\} + 24E_D\left\{e\right\}E_D\left\{e^2\right\}\nabla_{\mathbf{W}(k)} E_D\left\{e\right\} \\ & +12\left(E_D\left\{e\right\}\right)^2\nabla_{\mathbf{W}(k)} E_D\left\{e^2\right\} \\ & -24\left(E_D\left\{e\right\}\right)^3\nabla_{\mathbf{W}(k)} E_D\left\{e\right\})\end{aligned} \tag{3.52}$$

In this section, a number of numerical simulation results are presented for validating the arguments discussed in this chapter. The state-of-the-art generalization enhancement methods, including weight decay and weight elimination, as well as the two proposed higher-order statistics based objective functions have been studied in the prediction of two chaotic time-series, namely Henon Attractor and Sunspot time-series. The simulations were conducted under a SUN Sparc 20 platform.

3.2.4.1. Experiment one: Henon Attractor

The Henon map, which exhibits stochastic dynamics in the classical case, is a nonlinear quantum harmonic oscillator with two degrees of freedom, and is characterized by

$$x_{k+1} = 1 - ax_k^2 + bx_{k-1} \tag{3.53}$$

The Henon attractor is chaotic for the parameter values $a = 1.4$ and $b = 0.3$. The capacity of the attractor was estimated to be around *1.27*, and the size of the largest Lyapunov exponent is about *0.408*. The normalized $\widetilde{x}_k$ and $\widetilde{x}_{k-1}$, by $\widetilde{x}_k = \dfrac{x_k - \overline{x}}{\sigma}$ where $\overline{x}$ and σ are the mean and variance of the data, are the inputs of the neural networks, and the $\widetilde{x}_{k+1} + n_{k+1}$ is the target output where n_{k+1} is additive noise perturbation with zero mean and standard deviation of 0.1. Two different distributions, namely uniform and Gaussian, of noise perturbation were studied. Two hundred training patterns and another two set of 200 different test patterns were used in this study. The simulations were conducted under the conditions of learning rate *0.1* and momentum factor *0.9*. The architecture of the neural network used in this study is two inputs, ten hidden neurons, and one output. Each neural network was initialized by the same set of weights and was trained by 30,000 iterations.

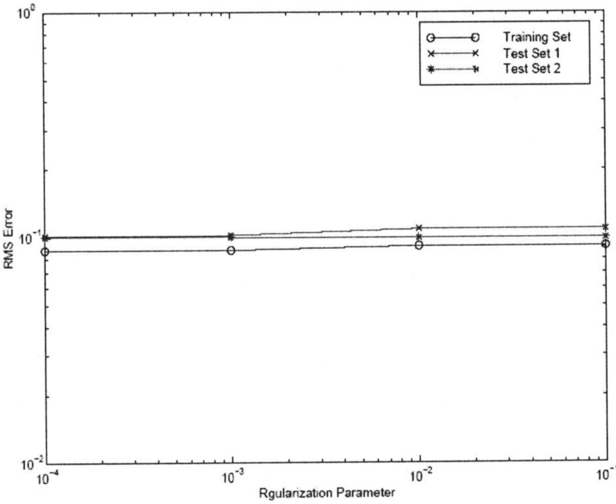

Figure 3.2. Results of Henon series prediction using HOC cost function under Gaussian noise perturbation

Under the noise perturbation of Gaussian distribution, different values of the regularization parameter λ were used to examine the effect to the network performance for the methods studied in this experiment. Tables 3.1 and 3.2 summarize the simulation results and Figs. 3.2 to 3.4 illustrate the variation in the network performance to the value of the regularization parameter λ. Fig. 3.2 manifests that varying the value of λ did not noticeably affect the network training and the network performance when the HOC objective function was applied. Figs. 3.3 and 3.4, respectively, show that the performance of the weight decay and weight elimination methods were significantly affected by the value of λ. Moreover, the methods are relatively sensitive to noise when a non-optimal λ is used. Additional simulation was conducted to validate the confidence upper bound of approximation error for the HOC objective function. The network trained according to $\lambda = 10^{-4}$ was tested. Another 45 test sets of 200 samples were used for validation. The RMS error of the 45 test set was computed and the histogram of the RMS errors is shown in Fig. 3.5. According to equation 3.35, we consider that, with a confidence level 99%, the confidence upper bound of the approximation error in the form of RMS error is *0.1344*. The upper bound is bounded above all the test errors as shown in Fig. 3.5. Hence, the approximation error can be coarsely estimated based on the upper bound when the HOC objective function is used and the noise perturbation is Gaussian distributed.

Table 3.1. Simulation results of Henon series prediction using the HOC objective functions under Gaussian noise perturbation. "Not Conv." means that the networks training does not converge to a training error lower than 1

	RMS Error		
λ	Training Set	Test Set 1	Test Set 2
10^{-4}	0.087690	0.10202	0.10015
10^{-3}	0.087668	0.10210	0.10013
10^{-2}	0.091421	0.10855	0.099337
0.1	0.091461	0.10848	0.099202
1	Not Conv.		
10	Not Conv.		

Table 3.2. Simulation results of Henon series prediction using the weight decay and weight elimination method under Gaussian noise perturbation. "Not Conv." means that the networks training does not converge to a training error lower than 1

	RMS Error		
λ	Training Set	Test Set 1	Test Set 2
10^{-4}	0.093319	0.10375	0.10127
10^{-3}	0.15442	0.15637	0.15520
10^{-2}	0.36620	0.36434	0.36155
0.1	Not Conv.		
1	Not Conv.		
10	Not Conv.		

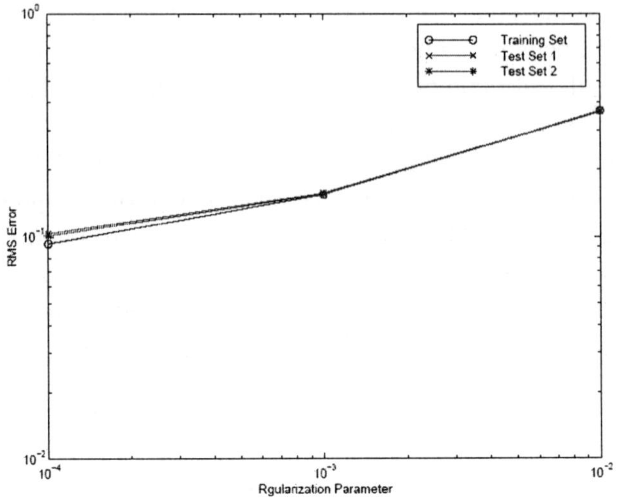

Figure 3.3. Results of Henon series predication using weight decay method under Gaussian noise perturbation

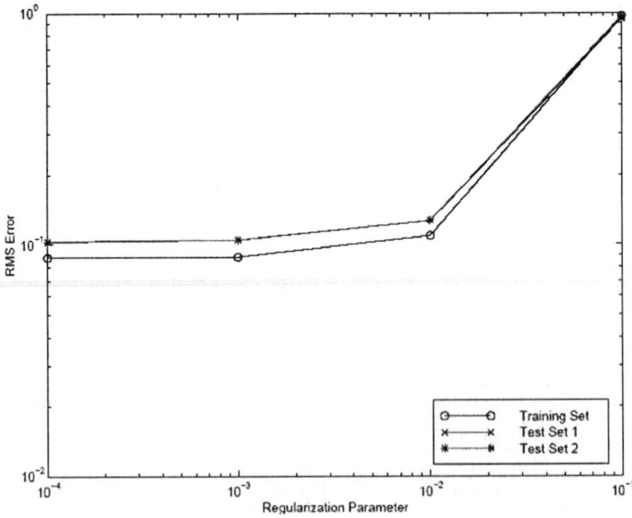

Figure 3.4. Results of Henon series predication using weight elimination method under Gaussian noise perturbation

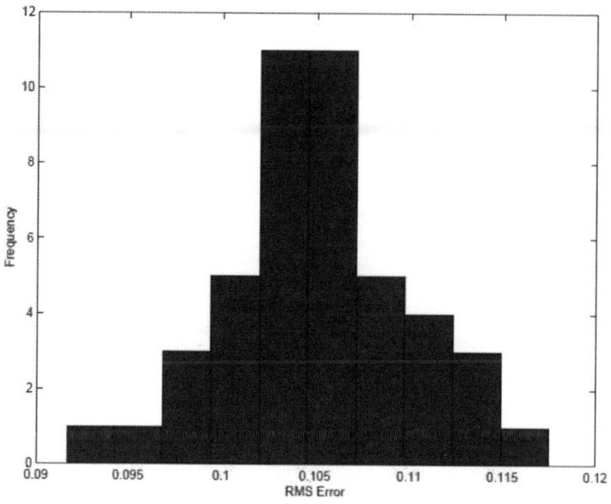

Figure 3.5. Histogram of the test error distribution using HOC cost function

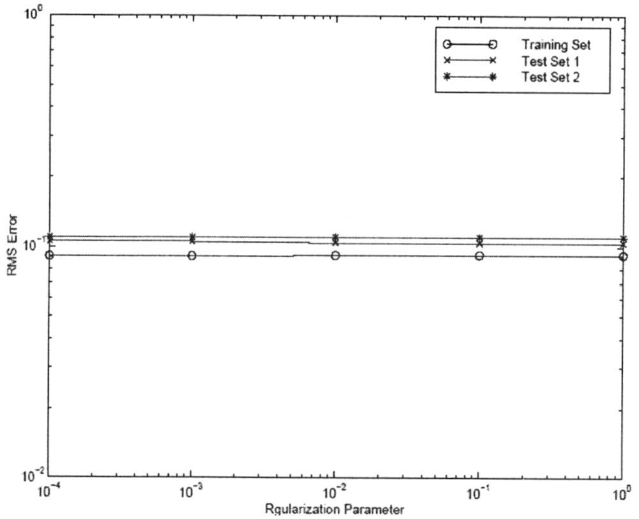

Figure 3.6. Results of Henon series prediction using HOC cost function under uniform noise perturbation

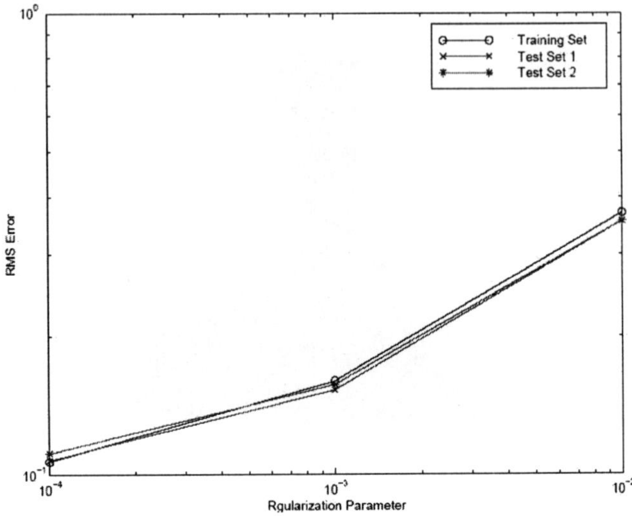

Figure 3.7. Results of Henon series prediction using weight decay method under uniform noise perturbation

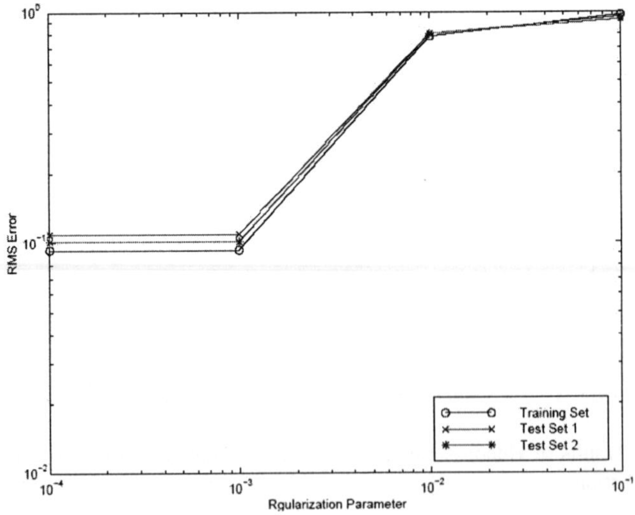

Figure 3.8. Results of Henon series prediction using weight elimination method under uniform noise perturbation

Under the noise perturbation of uniform distribution, similarly, different values of the regularization parameter λ were used to examine the effect on the network performance for the methods. Figs. 3.6 to 3.8 illustrate the variation in the network performances to the value of the regularization parameter λ. Fig. 3.6 illustrates that the network training and the network performance when the HOC objective function was applied. Figs. 3.7 and 3.8, respectively, show that the performance of the weight decay and weight elimination methods were significantly affected by the value of λ. Moreover, all the training errors and test errors in 3.3 are not noticeably deviated from the value of standard deviation of the noise perturbation. When the proper value of λ is selected, the weight decay and weight elimination methods can train a network to generalize the training data. The methods are relatively sensitive to noise when a non-optimal λ is used.

Furthermore, in accordance with the two sets of simulations under different noise perturbation, the results indicate that the performance of the HOC objective function and the weight decay method were not noticeably changed when different noise distributions. However, the performance of weight elimination method was significantly affected by the noise distribution. The HOC based objective function provides significantly better performance in terms of the generalization capability, the robustness to λ and noise perturbation.

3.2.4.2. Experiment Two: Sunspot time-series

The prediction of the sunspot series is regarded as a benchmark test for the time-series prediction technique. The sunspot data (1700-1979) are divided into a training set (1700-1920) and two test sets, covering the periods of 1921-1955 and 1956-1979. The network architecture is identical to that used by Weigend *et al.* (1991), which has 12 inputs, 8 hidden units, and 1 output. A thorough comparison amongst the weight decay method, the weight elimination method, and the FOC objective function are included hereafter. The learning rate of 0.1 and the momentum factor of 0.9 were used and each network was trained by 30,000 iterations. Different regularization parameters were applied to evaluate the performance of the generalization enhancement methods studied in this chapter. Figs. 3.9 to 3.11 illustrate the variation in the network performance to the value of the regularization parameter λ. Fig. 3.9 manifests that varying the value of λ does not noticeably affect the network training and the network performance when the HOC based objective function was applied. Figs. 3.10 and 3.11 show that the performance of the weight decay and weight elimination methods were significantly affected by the value of λ. Figs. 3.10 and 3.11 indicate that the generalization capabilities of the networks in terms of RMS errors of the test sets 1 and 2 were fluctuated enormously. These results manifest that it is relatively difficult to tune the generalization capability by varying λ when the weight decay and weight elimination methods are used. The networks trained by the weight decay and weight elimination methods are probably under-trained and the methods are relatively sensitive to the noise when a non-optimal λ is used.

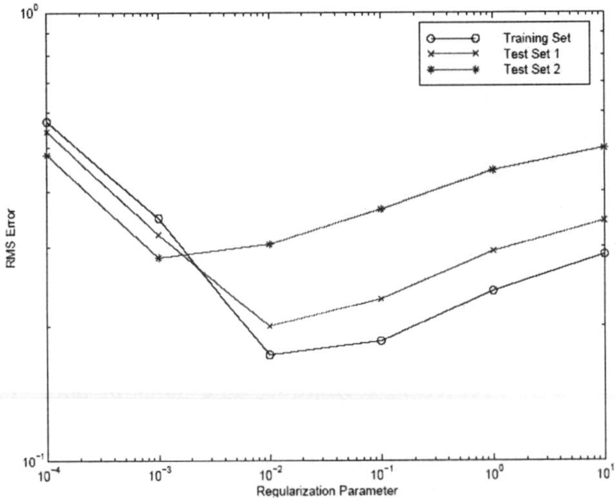

Figure 3.10. Sunspot series prediction using weight decay method

3.3. Regularization for Generalization Enhancement

In the Section 3.2, a new type of objective functions is introduced according to higher-order statistics. Basically, the HOC based objective function is constructed by the regularization technique. The technique constructs a regularized objective function to assimilate the *a priori* knowledge. For example, in the applications of classification, new discriminant functions (Setiono, 1997) were proposed to maximize the classification accuracy to the unseen examples. For the applications of functional approximation, a number of regularized objective functions have been proposed to enhance the generalization capability. For instance, the regularized objective functions are derived in accordance with the techniques such as searching flat minima (Hochreiter & Schmidhuber, 1997), minimizing the mutual information criterion (Deco et al., 1995), and minimizing the higher-order cumulants between the network outputs and the desired outputs. Despite the regularization technique being a systematic approach to make the network training less ill-posed, the training process may stall because of the existence of sub-optimal solutions.

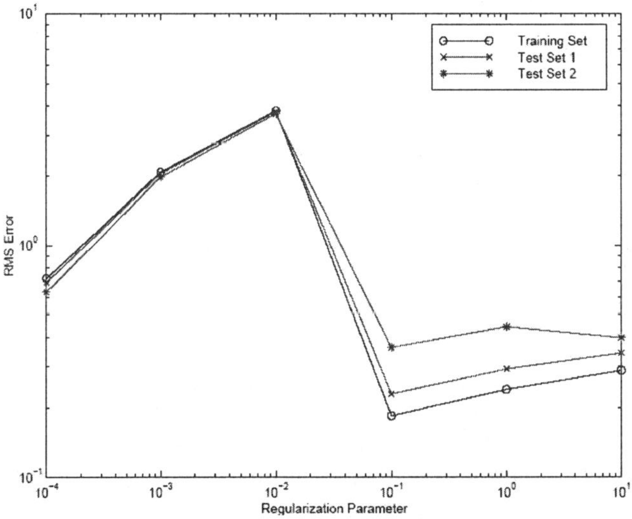

Figure 3.11. Sunspot series prediction using weight elimination method

We consider a typical form of the regularized objective function expressed in the following equation:

$$H(\mathbf{W},D)=M(\mathbf{W},D)+\lambda P(\mathbf{W},D) \tag{3.54}$$

where $\mathbf{W} = (w_0, w_1, \ldots, w_m)^T$ is the weight vector of the neural network and D is the set of training examples; and λ is the regularization parameter. $M(\mathbf{W}, D)$, which is mostly in terms of the LS error or the sum-of-squares error, is the primary cost term; and $P(\mathbf{W}, D)$ is the regularization term used to assimilate the *a priori* knowledge. For example, when weight decay method is used, the regularization term $P(\mathbf{W}, D)$ will be $\sum_i w_i^2$. In general, the plausible range of λ is often data dependent and has to be determined experimentally. The value of λ is often pre-selected within its plausible range, or it is selected in accordance with some heuristic selection schemes. It is believed that a systematic λ selection mechanism may be able to further enhance the generalization capability of the trained neural networks. In fact, this type of objective functions is still suffering from the problem of the existence of sub-optimal solutions because of the nonlinearity of neural networks. Although some undesired solutions should have been screened out to some extent, the regularized objective function introduces another set of undesired solutions. Consequently, the enhancement in the regularization technique may sometimes be insignificant, especially when a fixed value of λ is used during the network training.

From equation 3.54, the network training stalls only when $\nabla H(\mathbf{W}, D) = 0$.

The $\nabla M(\mathbf{W}, D)$ and $\nabla P(\mathbf{W}, D)$ are both zero vectors;

The $\nabla M(\mathbf{W}, D)$ and $\nabla P(\mathbf{W}, D)$ are both non-zero vectors such that

$$\nabla M(\mathbf{W}, D) + \lambda \nabla P(\mathbf{W}, D) = 0 \tag{3.55}$$

Condition 1 is a trivial case. The network training is often expected to converge to the minimum of this condition. Condition 2 may contribute to the introduction of another set of undesired sub-optimal solutions. The location of the sub-optimal solutions defined in Condition 2 is significantly affected by the pre-selected value of λ. Hence, the selection of λ is one of the major issues in the regularization technique and is determinant in the performance of neural networks, especially in the generalization capability. In this chapter, an adaptive regularization parameter selection method is introduced to tackle this problem.

Although the sub-optimal solutions defined in Condition 2 can be "bypassed" by adaptively changing the regularization parameter, there may exist sub-optimal solutions of Condition 1 because of the non-quadratic objective function. The sub-optimal solutions of Condition 1 are unable to be "bypassed" because the condition does not depend upon the value of the regularization parameter. The conventional gradient descent type optimizations, such as conjugate-gradient and quasi-Newton methods (Battiti, 1992), do not guarantee the convergence to an optimal solution because the gradient descent type algorithms are coarsely classified as local search methods which are based on the local information, namely gradient and Hessian matrix. Despite the local search approach being the most efficient and popular techniques, the network weights are easily trapped in a sub-optimal solution during the training process. Consequently, the properties of the universal approximator cannot be fully exploited and only a sub-optimal time-series predictor is obtained.

Another approach called Global Search methods, such as simulated annealing (Aarts & Korst, 1989) and genetic algorithm (Dodd, 1990), have been proposed to search global minima over the whole error surface. The Global Search approach is based on more information from the terrain of the error surface. Whether the global search methods are probabilistic or deterministic, the methods do not suffer from the problem of trapping in sub-optimal solutions. The global search methods introduce jumps when the neural network is stuck in a sub-optimal solution. To search the global minima in the extremely high-dimensional weight space is an extremely time-consuming procedure. Apart from local information, assimilating more terrain information of the error surface in the training process is very difficult without sacrificing the training time and increasing the computation complexity. Furthermore, the two discussed approaches are aimed at general applications and objective functions.

3.3.1. Adaptive Regularization Parameter Selection (ARPS) Method

As the selection of λ is extremely crucial to the performance of neural networks, this section introduces an adaptive λ selection mechanism to tackle the problems due to the sub-optimal solutions defined in Condition 2. The ARPS method consists of three main elements according to their functions. The three functional elements are responsible to the following functions:

Stalling Identification method identifies whether the training process converges to a sub-optimal solution that satisfies Condition 2;

λ Selection Scheme A selects an appropriate value of λ to ensure the training convergence of the $M(\mathbf{W},D)$ and $P(\mathbf{W},D)$ when the training process is not stuck at a sub-optimal solution that satisfies Condition 2;

λ Selection Scheme B selects an appropriate value of λ to ensure the training convergence of the $M(\mathbf{W},D)$ when the training process may stall in the sub-optimal solution.

On one hand, based on their functions, the λ selection scheme A guarantees the convergence of the $M(\mathbf{W},D)$ and $P(\mathbf{W},D)$ terms when there is no clue indicating the training process stalling. This part assures that the training process goes as smooth as possible. On the other hand, the λ selection scheme B will be applied to avoid the network training from stalling at a sub-optimal solution defined in Condition 2 when the ARPS method identifies the training process is about to stall at the sub-optimal solution. Hence, within the plausible range of the λ, the ARPS method is capable of avoiding the training process from stalling at a sub-optimal solution defined in Condition 2. The detailed description of the stalling identification method and the two λ selection schemes are given in the sections beneath.

3.3.1.1. Stalling Identification Method

From equation 3.55, the stalling situation of condition 2 occurs when the vector sum of the nonzero $\nabla M(\mathbf{W},D)$ and $\nabla P(\mathbf{W},D)$ terms are zero vector. This implies that the $\nabla M(\mathbf{W},D)$ and $\nabla P(\mathbf{W},D)$ are scalar multiple of each other, that is,

$$\nabla M(\mathbf{W},D) = -\lambda \nabla P(\mathbf{W},D) \tag{3.56}$$

Thus, Condition 2 can be easily identified by means of inner product of the direction vectors of the two gradient terms. The direction vector of a vector $\mathbf{x} = (x_1, x_2, \cdots, x_n)^T$ is defined by

$$\hat{\mathbf{x}} = \frac{\mathbf{x}}{\|\mathbf{x}\|} \tag{3.57}$$

and the inner product between a direction vectors $\hat{\mathbf{x}} = (\hat{x}_1, \hat{x}_2, \cdots, \hat{x}_n)^T$ and $\hat{\mathbf{y}} = (\hat{y}_1, \hat{y}_2, \cdots, \hat{y}_n)^T$ is defined by

$$\langle \hat{\mathbf{x}}, \hat{\mathbf{y}} \rangle = \sum_{i=1}^{n} \hat{x}_i \hat{y}_i \tag{3.58}$$

where the norm $\|\mathbf{x}\|$ is defined by

$$\|\mathbf{x}\| = \sqrt{\sum_{i=1}^{n} x_i^2} \tag{3.59}$$

Hence, the value of the inner product signifies the likelihood of getting stuck at a sub-optimal solution defined in Condition 2. In this project, the criterion of the stalling identification method is based on the value of the inner product $\langle \nabla \hat{M}, \nabla \hat{P} \rangle$. When the inner product is close to negative one, the training process almost stalls at a sub-optimal solution defined in Condition 2. Consequently, the mechanism of the stalling identification method is that the training process is classified as stalling when the inner product is less than a pre-selected threshold γ; otherwise, the training process is considered as not stalling.

3.3.1.2. λ Selection Schemes

Apart from the stalling identification method, the λ selection schemes are another of paramount importance in the ARPS method. The rationale behind the λ selection schemes is that when the training process is classified as not stalling, λ is selected to guarantee the convergence of the both $M(\mathbf{W}, D)$ and $P(\mathbf{W}, D)$ to maximize the effect of the regularization method. While the network training is about to stall, another λ is chosen to assure the convergence of the $M(\mathbf{W}, D)$ and $P(\mathbf{W}, D)$ only. The $P(\mathbf{W}, D)$ may not further converge, or even diverge slightly. In other words, the ARPS method, on one hand, breaks the tendency from getting stuck in the sub-optimal solution defined in Condition 2 by means of changing the λ. On the other hand, all the sub-optimal solutions defined in Condition 2 disappear momentarily because the training process, at that instance, is switched into a non-regularized training process. Consequently, the network training may tunnel through the sub-optimal solution.

In order to implement the above ideas, a set of λ selection criteria are derived and are obtained by means of the convergence analysis for the gradient descent type optimization. We let the plausible range of the λ for a particular regularized objective function be the interval $(\lambda_{\min},\lambda_{\max})$. Suppose the value of the inner product $\langle\nabla\hat{M},\nabla\hat{P}\rangle$ is negative. We consider the sufficient condition for the convergence of the $M(\mathbf{W},D)$ term. The change of $M(\mathbf{W},D)$ is given by

$$\Delta M = M(\mathbf{W}+\Delta\mathbf{W},D)-M(\mathbf{W},D) \tag{3.60}$$

Because the gradient descent type training technique is used in this project, the update step $\Delta\mathbf{W}$ is proportional to ∇H, viz. $\Delta\mathbf{W}=-\eta\nabla H$ where η is the learning rate. Using Taylor expansion, we have

$$\Delta M = \langle\nabla M,-\eta\nabla H\rangle \tag{3.61}$$

Using Lyapunov method, the sufficient condition for the convergence of $M(\mathbf{W},D)$ is given by

$$0<\frac{-\|\nabla M\|^2}{\langle\nabla M,\nabla P\rangle}\le\lambda \tag{3.62}$$

Similarly, the sufficient condition for the convergence of the term $P(\mathbf{W},D)$ is

$$0<\frac{-\langle\nabla M,\nabla P\rangle}{\|\nabla P\|^2}\le\lambda \tag{3.63}$$

The detailed derivation of the above sufficient conditions is summarized in Appendix 3.3. In order to guarantee the convergences of the both terms $M(\mathbf{W},D)$ and $P(\mathbf{W},D)$, the λ can be chosen within the following interval

$$0 < \max\left\{\frac{-\langle \nabla M, \nabla P\rangle}{\|\nabla P\|^2}, \frac{-\|\nabla M\|^2}{\langle \nabla M, \nabla P\rangle}\right\} < \lambda \tag{3.64}$$

when the inner product $\langle \nabla M, \nabla P\rangle$ is negative and greater than γ. When the inner product is greater than zero, the λ_A is set to be half of $\lambda_{\max}$ because a positive λ assures the convergence. To assure the convergence of $M(\mathbf{W}, D)$, the value of λ can be selected from the interval beneath:

$$0 < \lambda \leq \min\left\{\frac{-\langle \nabla M, \nabla P\rangle}{\|\nabla P\|^2}, \frac{-\|\nabla M\|^2}{\langle \nabla M, \nabla P\rangle}\right\} \tag{3.65}$$

In the λ selection scheme A, the λ is computed by

$$\lambda_A = \min\left\{\max\left\{\frac{-\langle \nabla M, \nabla P\rangle}{\|\nabla P\|^2}, \frac{-\|\nabla M\|^2}{\langle \nabla M, \nabla P\rangle}\right\}, \lambda_{\max}\right\} \tag{3.66}$$

when the value of the inner product $\langle \nabla \hat{M}, \nabla \hat{P}\rangle$ is negative. Otherwise, the value of λ remains unchanged. In the scheme B, the λ is calculated by

$$\lambda_B = \frac{1}{2}\left(\max\{0, \lambda_{\min}\} + \min\left\{\frac{-\langle \nabla M, \nabla P\rangle}{\|\nabla P\|^2}, \frac{-\|\nabla M\|^2}{\langle \nabla M, \nabla P\rangle}\right\}\right) \tag{3.67}$$

Once the ARPS method is applied, the advantages of the regularization method are maximized, and the problem of the sub-optimal solutions defined in Condition 2 is eliminated within the plausible range of λ. Hereafter, the algorithm outline of our ARPS method is summarized as follows:

To initialize $\mathbf{W}_0$ and λ_0;

$\mathbf{W}_{k+1} = \mathbf{W}_k + \Delta\mathbf{W}_k$;

If the training error is smaller than the presumed value, then stop;

If the training process is likely to stall according to the stalling identification method, then jump to step 6;

To select λ_{k+1}, based on λ selection scheme A, and jump to step 2;

To select λ_{k+1}, based on λ selection scheme B, and jump to step 2;

3.3.1. *Synthetic Function Mapping*

The results of applying the HOC objective functions defined in this chapter are described as to a synthetically generated data set by the following trigonometric function

$$y = \tanh\left(4x + \frac{\pi}{2}\right) - \tanh\left(4x - \frac{\pi}{2}\right) - 0.925 \tag{3.68}$$

The independent variable x was uniformly generated on the range from -1 to 1. Two collections of the data were generated. The training set possesses 200 samples and the test set has another 200 samples. The property of the HOC based objective function is validated by applying the objective functions with ARPS method to approximate the synthetic function. Ten Monte Carlo runs were performed. The simulation results are tabulated in Table 3.3. It manifests that the network training is able to converge to a very low training error and comparably low test error. In addition, the deviation in the errors over the Monte Carlo runs is rather small as compared to the training errors. Fig. 3.12 illustrates the convergence curves of the network training. From the figures, it is observed that there is no observable sign showing the training being stuck and a tendency to converge to a lower training error. This result indirectly substantiates that the network training using HOC based objective function with ARPS method does not get stuck at sub-optimal solutions of Conditions 1 and 2. Figs. 3.13 and 3.14 illustrate the distributions of the magnitude of the weights of the neural network in the 10 runs. Fig. 3.13 shows the distributions of the sorted magnitude of the hidden layer weights in the 10 runs. The result indicates that there are only two dominant weights. These results are inline with the synthetic function which can be approximated by a neural network with 2 hidden neurons. Fig. 3.14 illustrates the distribution of the sorted magnitude of the output layer biases in the 10 runs. The figure indicates that the magnitude of the output layer biases in the 10 runs are close to the constant term in equation 3.68. These simulation results substantiate that the network training based on the HOC objective function with ARPS method is not trapped in sub-optimal solutions.

Table 3.3. Simulation results of synthetic function mapping

	RMS Error	
	Training Set	Testing Set
mean	0.0148	0.0155
standard deviation	0.0016	0.0016

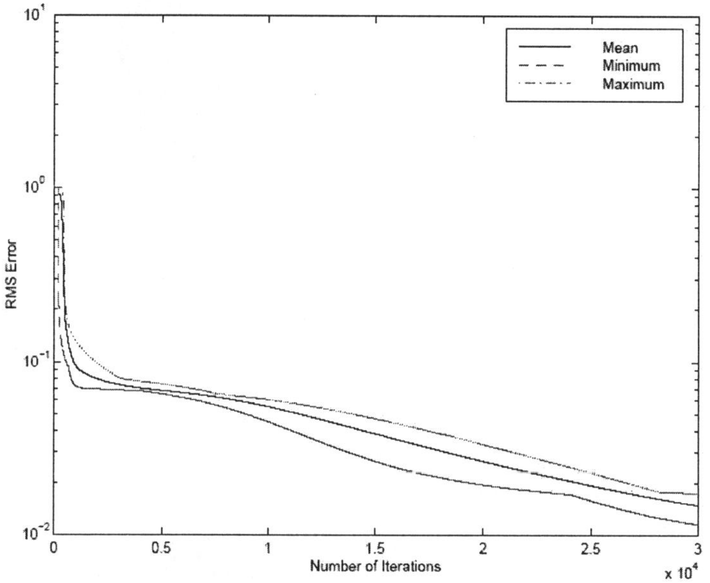

Figure 3.12. Convergence of the network training using HOC cost function. The upper curve is the maximum values of the training error over the 10 Monte Carlo runs. The middle curve is the mean values, and the lower curve is the minimum values

3.4. Concluding Remarks

This chapter addressed the problem of the generalization capability when only a finite number of observable data is available. Since neural networks are considered as universal approximators, the network training is in fact a nonparametric estimation. The training based on finite observable data is ill-posed. In other words, the trained neural network may memorize the data set rather than generalize it. This chapter has shown how the methods of objective function selection are applied to enhance the generalization capability.

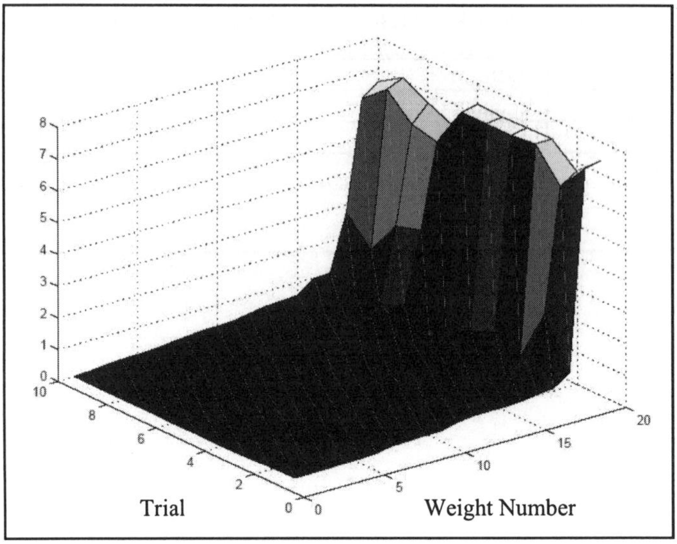

Figure 3.13. The distribution of the sorted magnitude of the hidden layer weights using HOC cost function with APRS method

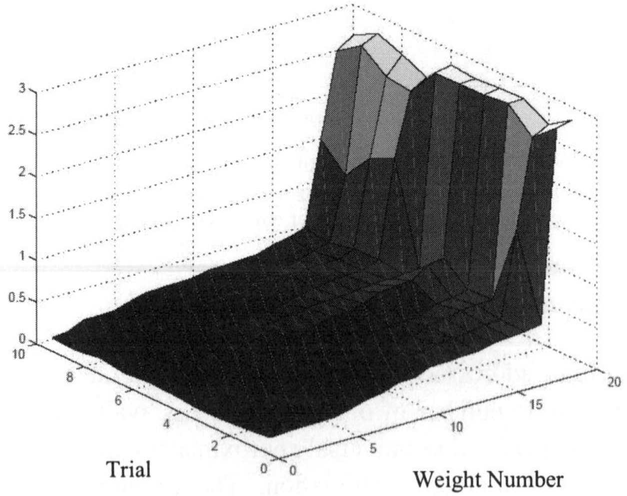

Figure 3.14. The distribution of the sorted magnitude of the output layer weights using HOC cost function with APRS method

3.4.1. *Objective function selection*

The least squares (LS) error function or mean squared error function is used to find maximum likelihood estimation assuming that the noise perturbation or the target value of the training data are Gaussian normal distributed. The sum-of-squares error function is only in terms of the second-order moment which cannot distinguish between the Gaussian distribution and other distributions having the same mean and variance. Thus, on one hand, the distribution of the residual error, $e_k = t_k - F(\mathbf{x}_k, \mathbf{W})$, of the trained network may not be Gaussian distributed because the sum-of-squares error function is in terms of the second-order moment. This degrades the generalization capability of the trained neural network. On the other hand, the distribution of real world data is non-Gaussian in general. The network training based on the LS error function is not able to determine a maximum likelihood estimate and results in a biased neural network because of improper assumption.

In this chapter, higher-order statistics based objective functions were described. The fourth-order cumulant (FOC) objective function is constructed by including a fourth-order cumulant regularization term in the mean squared error function. A maximum likelihood estimate can be obtained by applying the FOC objective function when the output variable or the noise perturbation is Gaussian distributed. Also, the signal-to-noise ratio is boosted because the cumulant term is zero when the noise perturbation is Gaussian distributed. The inclusion of the fourth-order cumulant term is capable of facilitating the network training because the term can extract more high-order information from the training data. To include the fourth-order cumulant term also enables the network training to capture the phase information from the data which is extremely essential to time-series forecasting.

The HOC objective function is capable of not only characterizing a distribution of non-Gaussianity but also measuring the kurtosis or "Gaussianity" of the distribution. Consequently, the FOC objective function enables the network training to filter out the undesired functional estimates of which the residual error is non-Gaussian distributed. The residual error is squeezed to be Gaussian distributed as much as possible. Thus, the trained network is more likely to have a high generalization capability when the output of the target system or the noise perturbation is Gaussian distributed. The distribution of the residual error of the trained network is very close to Gaussian. Therefore, the reliability of the network training in terms of the confidence upper bound of the approximation error can be coarsely estimated based on the test error of the novice data set.

In real world data, an additive noise may not be Gaussian distributed as the assumption of the sum-of-squares error function and the FOC objective function. Depending upon the working conditions of the target system, the additive noise is often symmetrically distributed. The computation complexity of the FOC objective function based algorithm is relatively high despite the advantages of the FOC objective function. This high computation complexity limits the variety of applications. The FOC objective function is capable of not only characterizing a non-Gaussian distribution but also measuring the skewness of the distribution. As a result, the FOC objective function enables the network training to filter out the undesired functional estimates of which the residual error is asymmetrically distributed. The residual error is squeezed to be symmetrically distributed as much as possible. Thus, the trained network may have a higher generalization capability when the output of the target system or the noise perturbation is symmetrically distributed. Moreover, an interesting property between $E\{e^2\}$ and $Cum\{e^r\}$, $r=3,4$ is found when the neural network is applied to time-series forecasting. If there exists a $\mathbf{W}$ such that, $\nabla E_k\{e_k^2\}=0$ and $\nabla Cum_k\{e_k^2\}=0$ for $r=3,4$ then

$$E_k\left\{\left(F(\mathbf{x}_k,\mathbf{W}^*)-F(\mathbf{x}_k,\mathbf{W})\right)^2\right\}=0.$$

Based on the property, new HOC basedobjective functions are reconstructed such that there is no sub-optimal solution defined in Condition 1. Consequently, we can say that FOC and TOC objective functions based algorithms are capable of ensuring that the network training is not trapped in sub-optimal solutions.

3.4.2. Regularization selection

It is well-known that the training process based on a regularized objective function may stall because of the existence of sub-optimal solutions. The conditions of getting stuck at sub-optimal solutions are summarized as follows:

The gradients of the primary and regularization terms are both zero vectors;

The gradients of the primary $M(\mathbf{W},D)$ and regularization $P(\mathbf{W},D)$ terms are both non-zero vectors such that

$$\nabla M(\mathbf{W},D)+\lambda\nabla P(\mathbf{W},D)=0.$$

In this chapter, the adaptive regularization parameter selection method was described to tackle this problem of getting stuck at sub-optimal solutions as defined in Condition 2. The methodology of the ARPS method is that when the training process is classified as not stalling, a λ is selected to guarantee the convergence of the both $M(\mathbf{W},D)$ and $P(\mathbf{W},D)$ to maximize the effect of the regularization method. While the network training is about to stall, another λ is chosen to assure the convergence of the $M(\mathbf{W},D)$ only. The $P(\mathbf{W},D)$ may not further converge, or even diverge slightly. In other words, the ARPS method, on one hand, breaks the tendency to getting stuck in the sub-optimal solution as defined in Condition 2 by means of changing the λ. On the other hand, all the sub-optimal solutions as defined in Condition 2 disappear momentarily because the training process is, at that instance, switched into a non-regularized type training. Consequently, the network training may tunnel through the sub-optimal solution and the advantages of the regularization method are maximized and the problem of the sub-optimal solution of Condition 2 is eliminated within the plausible range of λ. The ARPS method can only tackle the problem of sub-optimal solutions as defined in Condition 2. In general, there may exist sub-optimal solutions as defined in Condition 1.

Appendix 3.1: Confidence Upper Bound of Approximation Error

For the sake of simplicity, a multiple-input-single-output and time-invariant system is analyzed. The generalization to a multiple-input-multiple-output system is obvious. Suppose the m-input-single-output unknown target function can be summarized by a function $h:\mathbf{X}\mapsto\mathbf{Y}$ where the domain $\mathbf{X}$ is contained in $\Re^m$ and the range $\mathbf{Y}$ is a subset of $\Re$. An n-input-single-output neural network $F(\mathbf{x},\mathbf{W})$ with sufficiently large network size is applied. Because the measurement of system output is always perturbed, the target output t can be written as

$$t = h(\mathbf{x}) + n \tag{A3.1}$$

where n is a random noise which is Gaussian distributed, and is independent to $h(\mathbf{x})$ and the past t. The network training is aimed at finding the optimal weights $\mathbf{W}^*$ for an neural network model class $F(\mathbf{x},\mathbf{W})$ based on the minimization of the approximation error, i.e.

$$\varepsilon_a = E\left\{\left(t - F\left(\mathbf{x},\mathbf{W}^*\right)\right)^2\right\} \tag{A3.2}$$

A finite number of test data $D_t = \{(\mathbf{x}_i, t_i)\}, i = 1,2,\cdots,n$ is generally sampled from the target system so that ε_a can be estimated based on the finite number of samples. Test error ε_t, which is the estimate of ε_a, is defined as

$$\varepsilon_t = \frac{1}{n}\sum_{i=1}^{n}\left(t - F\left(\mathbf{x}_i,\mathbf{W}^*\right)\right)^2 \tag{A3.3}$$

When the FOC objective function is applied in the network training, the residual error $e = t - F(\mathbf{x},\mathbf{W})$ of the trained network should be squeezed to be as Gaussian distributed as possible. Suppose we obtain a test data set D_t which is new to the trained network $F(\mathbf{x},\mathbf{W}^*)$. It is assumed that the neural network $F(\mathbf{x},\mathbf{W})$ is sufficiently general to approximate the unknown target system $h(\mathbf{x})$, i.e. $h(\mathbf{x}) \in \{F(\mathbf{x},\mathbf{W})\}$; and the residual error e_i is independent and identically Gaussian distributed with variance σ^2, i.e. $e_i \sim N(0,\sigma^2)$. Consequently, we have, from equation A3.2,

$$\varepsilon_a = E\{e^2\} = \sigma^2 \tag{A3.4}$$

Let the statistic

$$s^2 = \frac{\sum_{i=1}^{n}(e_i - \bar{e})^2}{n-1} \tag{A3.5}$$

which is an unbiased estimator for σ^2. Since $\frac{(n-1)s^2}{\sigma^2}$ is distributed as chi-square distribution with *n-1* degree of freedom, i.e. $\frac{(n-1)s^2}{\sigma^2} \sim \chi^2_{n-1}$, with confidence level 100β%,

$$\frac{(n-1)s^2}{\sigma^2} \geq \chi^2_{n-1}(1-\beta) \tag{A3.6}$$

where

$$\chi^2_n(x) = \int_x^{\infty} \frac{1}{2^{n/2}\Gamma(n/2)} y^{\frac{n}{2}-1} \exp(-y/2)\,dy \tag{A3.7}$$

Subsequently, we have

$$\sigma^2 \leq \frac{(n-1)s^2}{\chi^2_{n-1}(\beta)} \tag{A3.8}$$

According to the central limit theorem, $\bar{e}$ is negligible for large n because $\sigma^2 >> Var\{\bar{e}\} = \frac{\sigma^2}{n}$. s^2 can be approximated by

$$s^2 \approx \frac{1}{n}\sum_{i=1}^{n} e_i^2 = \frac{n\varepsilon_t}{n-1} \tag{A3.9}$$

As a result, the confidence upper bound of ε_a is

$$\varepsilon_a = \sigma^2 \leq \frac{n\varepsilon_t}{\chi^2_{n-1}(\beta)} \tag{A3.10}$$

Appendix 3.2: Proof of the Property of the HOC Cost Function

The proof of the theorem 3.1 is as follows:

Lemma 3.1. For all w in **W**, if $\frac{\partial E_k\{e_k^2\}}{\partial w}=0$, then $E_k\{e_{\mathbf{w}_k}\}=0$

Proof:

$$\frac{\partial E_k\{e_k^2\}}{\partial w^1}=-2E\left\{e_k\frac{\partial F(\mathbf{x}_k,\mathbf{W})}{\partial w^1}\right\}=-2\left(E_k\{e_{\mathbf{w}_k}\}+E_k\{e_k^*\}\right)=-2E_k\{e_{\mathbf{w}_k}\}.$$

Thus, if, for all w in **W**, $\frac{\partial E_k\{e_k^2\}}{\partial w}=0$, then $E_k\{e_{\mathbf{w}_k}\}=0$. Q.E.D.

Theorem 3.1. If there exists a **W** such that, for all w in **W** and r=*3,4*, $\frac{\partial E_k\{e_k^2\}}{\partial w}=0$ and $\frac{\partial Cum_k\{e_r^3\}}{\partial w}=0$, then

$$E_k\left\{\left(F(\mathbf{x}_k,\mathbf{W}^*)-F(\mathbf{x}_k,\mathbf{W})\right)^2\right\}=E_k\{e_{\mathbf{w}_k}^2\}=0$$

Proof:

Suppose there exists a **W** such that $E_k\{e_{\mathbf{w}_k}^2\}\neq 0$, and, for all w be an element of **W**, $\frac{\partial E_k\{e_k^2\}}{\partial w}=0$ and $\frac{\partial Cum_k\{e_{\mathbf{w}_k}^2\}}{\partial w}=0$. For r=*3*, we have, from H2,

$$Cum_k\{e_k^3\}=Cum_k\{e_{\mathbf{w}_k}^3\}+Cum_k\{(e_k^*)^3\}$$

Subsequently,

$$\frac{\partial Cum_k\{e_k^3\}}{\partial w} = \frac{\partial Cum_k\left\{(e_{\mathbf{w}_k})^3\right\}}{\partial w}$$
$$= \frac{\partial E_k\{e_{\mathbf{w}_k}^3\}}{\partial w} - 3\left(E_k\{e_{\mathbf{w}_k}\}\frac{\partial E_k\{e_{\mathbf{w}_k}^2\}}{\partial w} + E_k\{e_{\mathbf{w}_k}^2\}\frac{\partial E_k\{e_{\mathbf{w}_k}\}}{\partial w}\right)$$
$$-6\left(E_k\{e_{\mathbf{w}_k}\}\right)^2\frac{\partial E_k\{e_{\mathbf{w}_k}\}}{\partial w}$$

By Lemma 3.1, we have

$$\frac{\partial Cum_k\{e_k^3\}}{\partial w} = \frac{\partial E_k\{e_{\mathbf{w}_k}^3\}}{\partial w} - 3E_k\{e_{\mathbf{w}_k}^2\}\frac{\partial E_k\{e_{\mathbf{w}_k}\}}{\partial w}$$
$$= 3E_k\left\{e_{\mathbf{w}_k}^2\frac{\partial e_{\mathbf{w}_k}}{\partial w}\right\} - 3E_k\{e_{\mathbf{w}_k}^2\}E_k\left\{\frac{\partial e_{\mathbf{w}_k}}{\partial w}\right\}$$
$$= 3\left[E_k\left\{e_{\mathbf{w}_k}^2\frac{\partial e_{\mathbf{w}_k}}{\partial w}\right\} - E_k\{e_{\mathbf{w}_k}^2\}E_k\left\{\frac{\partial e_{\mathbf{w}_k}}{\partial w}\right\}\right]$$
$$= -3\left[E_k\left\{e_{\mathbf{w}_k}^2\frac{\partial F}{\partial w}\right\} - E_k\{e_{\mathbf{w}_k}^2\}E_k\left\{\frac{\partial F}{\partial w}\right\}\right]$$

Thus, since $\frac{\partial Cum_k\{e_k^3\}}{\partial w} = 0$, we find

$$E_k\left\{e_{\mathbf{w}_k}^2\frac{\partial F}{\partial w}\right\} = E_k\{e_{\mathbf{w}_k}^2\}E_k\left\{\frac{\partial F}{\partial w}\right\}$$

Therefore, $e_{\mathbf{w}_k}^2$ and $\frac{\partial F}{\partial w}$ are uncorrelated [Papoulis91] because

$$Cov\left\{e_{\mathbf{w}_k}^2, \frac{\partial F}{\partial w}\right\} = E_k\left\{e_{\mathbf{w}_k}^2\frac{\partial F}{\partial w}\right\} - E_k\{e_{\mathbf{w}_k}^2\}E_k\left\{\frac{\partial F}{\partial w}\right\} = 0$$

However, using Taylor's expansion,

$$F(\mathbf{x}_k, \mathbf{W}^*) \approx F(\mathbf{x}_k, \mathbf{W}) + (w^* - w)\frac{\partial F(\mathbf{x}_k, \mathbf{W})}{\partial w}$$

Then, we have

$$e_{\mathbf{w}_k}^2 \approx \left(F(\mathbf{x}_k, \mathbf{W}) + (w^* - w)\frac{\partial F(\mathbf{x}_k, \mathbf{W})}{\partial w} - F(\mathbf{x}_k, \mathbf{W}) \right)^2 \approx (w^* - w)\frac{\partial F(\mathbf{x}_k, \mathbf{W})}{\partial w}.$$

Therefore, $e_{\mathbf{w}_k}^2$ and $\frac{\partial F}{\partial w}$ should be correlated. Contradiction!!!

$E\left\{\left(F(\mathbf{x}_k, \mathbf{W}^*) - F(\mathbf{x}_k, \mathbf{W})\right)^2\right\} = 0$, when $\frac{\partial E_k\{e_k^2\}}{\partial w} = 0$ and $\frac{\partial Cum_k\{e_{\mathbf{w}_k}^2\}}{\partial w} = 0$.

Similarly, for $r=4$, we have, by Lemma 3.1,

$$\begin{aligned} \frac{\partial Cum_k\{e_k^4\}}{\partial w} = \frac{\partial Cum_k\{e_{\mathbf{w}_k}^4\}}{\partial w} &= -4\left(E_k\left\{ e_{\mathbf{w}_k}^3 \frac{\partial F}{\partial w} \right\} - E_k\left\{\frac{\partial F}{\partial w}\right\} E_k\left\{e_{\mathbf{w}_k}^3\right\} \right). \\ &+ 12 E_k\left\{e_{\mathbf{w}_k}^2\right\} E_k\left\{ e_{\mathbf{w}_k} \frac{\partial F}{\partial w} \right\} \end{aligned}$$

By Orthogonality principle, we obtain

$$\frac{\partial Cum_k\{e_k^4\}}{\partial w} = -4\left(E_k\left\{ e_{\mathbf{w}_k}^3 \frac{\partial F}{\partial w} \right\} - E_k\left\{\frac{\partial F}{\partial w}\right\} E_k\left\{e_{\mathbf{w}_k}^3\right\} \right).$$

Since $\frac{\partial Cum_k\{e_k^4\}}{\partial w} = 0$,

$$Cov\left\{ e_{\mathbf{w}_k}^3, \frac{\partial F}{\partial w} \right\} = E_k\left\{ e_{\mathbf{w}_k}^3 \frac{\partial F}{\partial w} \right\} - E_k\left\{e_{\mathbf{w}_k}^3\right\} E_k\left\{\frac{\partial F}{\partial w}\right\} = 0.$$

It implies $e_{\mathbf{w}_k}^3$ and $\frac{\partial F}{\partial w}$ are uncorrelated. But, according to Taylor expansion, $e_{\mathbf{w}_k}$ and $\frac{\partial F}{\partial w}$ are correlated and so does the case of $e_{\mathbf{w}_k}^2$ and $\frac{\partial F}{\partial w}$. Contradiction!!!

$E\left\{\left(F(\mathbf{x}_k,\mathbf{W}^*)-F(\mathbf{x}_k,\mathbf{W})\right)^2\right\}=0$, when $\frac{\partial E_k\left\{e_k^2\right\}}{\partial w}=0$ and $\frac{\partial Cum_k\left\{e_{\mathbf{w}_k}^4\right\}}{\partial w}=0$. Q.E.D.

Appendix 3.3: The Derivation of the Sufficient Conditions of the Regularization Parameter

We consider the regularized objective function is defined as

$$H(\mathbf{W})=M(\mathbf{W})+\lambda P(\mathbf{W}) \tag{A3.11}$$

where λ is the regularization parameter and is positive. In this derivation, a gradient descent type training method is considered, viz.

$$\Delta\mathbf{W}=-\eta\nabla H \tag{A3.12}$$

where η is the learning factor.

Now, we consider the convergence of $M(\mathbf{W})$. After each iteration, the change of $M(\mathbf{W})$ is given by

$$\Delta M=M(\mathbf{W}+\Delta\mathbf{W})-M(\mathbf{W}) \tag{A3.13}$$

Since $\Delta\mathbf{W}=-\eta\nabla H$, using Taylor expansion, we have

$$\begin{aligned}\Delta M &\approx \left(M(\mathbf{W})+\langle\nabla M(\mathbf{W}),\Delta\mathbf{W}\rangle\right)-M(\mathbf{W})\\ &\approx \langle\nabla M(\mathbf{W}),-\eta\nabla H(\mathbf{W})\rangle\\ &\approx -\eta\left(\langle\nabla M(\mathbf{W}),\nabla M(\mathbf{W})\rangle+\lambda\langle\nabla M(\mathbf{W}),\nabla P(\mathbf{W})\rangle\right)\end{aligned} \tag{A3.14}$$

Using Lyapunov method, we have $\Delta M\le 0$ and subsequently, we obtain the sufficient condition for the convergence of $M(\mathbf{W})$

$$\lambda \geq \frac{-\|\nabla M(\mathbf{W})\|^2}{\langle \nabla M(\mathbf{W}), \nabla P(\mathbf{W})\rangle} \tag{A3.15}$$

The convergence of $P(\mathbf{W})$ is now considered. Similar to the case of $M(\mathbf{W})$, we have

$$\begin{aligned}\Delta P &= P(\mathbf{W}+\Delta\mathbf{W}) - P(\mathbf{W}) \\ &\approx \langle \nabla P(\mathbf{W}), \Delta\mathbf{W}\rangle \\ &\approx -\eta\left(\langle \nabla P(\mathbf{W}), \nabla M(\mathbf{W})\rangle + \lambda\|\nabla P\|^2\right)\end{aligned} \tag{A3.16}$$

Consequently, the sufficient condition for the convergence of $P(\mathbf{W})$ is

$$\lambda \geq \frac{\langle \nabla P(\mathbf{W}), \nabla M(\mathbf{W})\rangle}{-\|\nabla P(\mathbf{W})\|^2} \tag{A3.17}$$

Exercises

Q3.1. Let $\mathbf{v} = col(v_1, v_2, \ldots, v_k)$ and $\mathbf{x} = col(x_1, x_2, \ldots, x_k)$, where $(x_1, x_2, \ldots, x_k)$ denotes a collection of random variables. The kth-order cumulant of these random variables is defined as the coefficient of $(v_1, v_2, \ldots, v_k) \times j^k$ in the Taylor series expansion (provided it exists) of the cumulant-generating function

$$K(\mathbf{v}) = \ln E\left\{\exp\left(j\mathbf{v}^T\mathbf{x}\right)\right\}$$

Prove that $cum(x_1, x_2) = E\{x_1 x_2\}$ using the above equation and the definition in Eq. (3.17).

Q3.2. Apply the definitions and properties stated in Section 3.2.2 and 3.2.3 to derive the third-order cumulant cost function and hence derive a batch-mode third-order cumulant cost function based backpropagation algorithm.

Q3.3. Consider the Fourth Order Cumulant cost function as shown in Eq. (3.45) as a regularized error function and suppose that the unregularized error E_D is minimized by a weight vector **w***. Show that, if the regularization coefficient λ is small, the weight vector $\tilde{\mathbf{w}}$ which minimizes the regularized error can be written in the form

$$\tilde{\mathbf{w}} = \mathbf{w}^* - \lambda \mathbf{H}^{-1} \nabla Cum_D \left\{ \left(t - F(\mathbf{x}, \mathbf{W}) \right)^4 \right\}$$

where the gradient $\nabla Cum_D \left\{ \left(t - F(\mathbf{x}, \mathbf{W}) \right)^4 \right\}$ and the Hessian $\nabla\nabla E_D$ are evaluated at $\mathbf{w} = \mathbf{w}^*$.

Q3.4. Use the derivation in Appendix 3.3 to derive the sufficient conditions of the regularization parameter λ in the Fourth Order Cumulant Objective Function as shown in Eq. (3.45). Assume that a gradient descent type training method is used and the $\Delta \mathbf{W} = -\eta \nabla H$ where η is the learning factor.

Chapter 4

Basis Function Networks for Classification

In the nervous system of biological organisms there is evidence of neurons whose response characteristics are "local" or "tuned" to some region of input space. An example is the orientation-sensitive cells of the visual cortex, whose response is sensitive to local regions in the retina. In this chapter, we discuss a network structure related to the multilayer feedforward network, known as the Radial Basis Function (RBF) network. The RBF network is a feedforward structure with a modified hidden layer and training algorithm, which may be used for mapping. The RBF network can also be seen as a class of classifiers. It is characterized by having a transfer function in hidden unit layer, having radial symmetry with respect to a centre. From this characteristic it comes the name RBF. Usually the function is also bell-shaped, thus the activation of the unit has a maximum in the center and is almost equal to zero far from it. This feature entails the possibility to modify a unit of the network without affecting the overall behaviour and turns out to be very useful in order to implement incremental learning strategies. Moreover, they exhibit nice mathematical properties that exploit the regularization theory, and they are suitable to statistical and symbolic interpretations.

The chapter is composed of three complementary parts. The first part (Section 4.1 and 4.2) gives the theoretical background related to basic function network for classification. The second part (Section 4.3) presents how the RBF networks work. The third part (Section 4.4) addresses the connections between RBF networks and other approaches. More in detail, Section 4.1 presents the basic theories of the linear separation and perceptions in which we can prove that two groups are linearly separable if there is a linear function. Section 4.2 gives some of the possible ways to describe more general functions of the feature variables which can be basis function models for parametric smoothing. Section 4.3 describes the basic architecture of RBF network and it approximation properties, i.e. the characterization of the problems that RBF networks can solve. Some common learning algorithms used for RBF networks are also presented in the same section. Consequently, the regularization theory of the RBF networks is discussed. Section 4.4 discusses RBF networks could be considered as some advance models, such as Support Vector Machine, Wavelet Network, Fuzzy RBF network and Probabilistic Network Networks, in which they share some common properties of the RBF networks. Finally, Section 4.5 draws the concluding remarks of this chapter.

4.1. Linear Separation and Perceptions

Historically, special attention has been given to situations in which two species are completely separated on the first linear discriminant. We always say two groups are linearly separable if there is a linear function of the variables, say $\mathbf{xa}+b$, which is positive on one group and negative on the other. A function which computes a linear combination of the variables and returns the sign of the result is known as a perceptron after the work of F. Rosenblatt (1957, 1958, 1962). There is also publication by (Block, 1962). Their interest now is in their continuing influence on the thinking in the field of neural networks.

Let us add a column of 1's to $\mathbf{x}$ and add b to $\mathbf{a}$. Let $\mathbf{z}=\mathbf{x}/\|\mathbf{x}\|$ on the first group and $\mathbf{z}=-\mathbf{x}/\|\mathbf{x}\|$ on the second group. We then seek a linear combination $\mathbf{a}$ such that $\mathbf{za}>\mathbf{0}$ for every example in the training set. Since the training set is finite, we can choose $\delta>0$ so that $\mathbf{za}>\delta$. Indeed, we can achieve this for any $\delta>0$ by rescaling $\mathbf{a}$.

One approach to the problem would be to choose $\mathbf{a}$ by least squares to make $\mathbf{za}$ as near one as possible, or to regress $y=\pm 1$ on $\mathbf{x}$ which as we have seen gives the linear discriminant up to a scale factor. However, there is no guarantee that the linear discriminant will linearly separate the groups if they are linearly separable, and it is easy to construct examples in which it will not. A more direct formulation is to minimize the number of errors, but as that is a discrete measure, the optimization is difficult. The sum of the degree of error $\sum_i [\delta - \mathbf{z}_i\mathbf{a}]_+$ will be zero if and only if linear separation can be achieved. This is equivalent to solving the linear programming problem $\mathbf{z}_i\mathbf{a} \geq \delta$, and linear programming methods can find a solution or show that none exists (Grinold, 1969).

In the late 1950s, a number of researchers were interested in simpler but iterative solutions, in which the value of **a** was adjusted after each example was presented. The derivative for the least-squares problem $\|y - Xa\|^2$ is $2X^T(Xa - y)$ and so a steepest descent procedure would be of the form

$$a \leftarrow a - \eta \sum_i (x_i a - y_i) x_i^T \tag{4.1}$$

For small amount of η, this process converges to the space of least-squares solutions.

Rather than compute the sum on the right-hand side and update **a**, we could update after each pattern was considered. This gives the rule

$$a \leftarrow a - \eta (x_i a - y_i) x_i^T \tag{4.2}$$

known as *Widrow-Hoff learning* (Widrow and Hoff, 1960) or the *delta rule*. The patterns are presented cyclically until convergence, which will need $\eta \to 0$.

Rosenblatt's perceptron learning rule replaced the term $x_i a$ in (4.2) by the output of the perceptron, the sign of **xa**. Thus, **a** is changed only if the current pattern is misclassified, and so the rule is of the form $a \leftarrow a + 2\eta z_i^T I(z_i a \leq 0)$. No generality is lost by taking $\eta = 1/2$, since we can rescale **a**. Rosenblatt showed that this rule will converge in a finite number of steps to a linearly separating combination if one exists. Let **a*** be a suitable combination chosen so that $z_i a^* \geq 1$ for all members of the training set. If the rule changes **a**, we have $z_i a^* \leq 0$ so

$$\|a + \Delta a - a^*\|^2 = \|a - a^*\|^2 + 2z_i(a - a^*) + 1 \leq \|a - a^*\|^2 - 1$$

This shows the rule terminates in at most $\|a_0 - a^*\|^2$ steps.

This result is known as the perceptron convergence theorem. Its limitations were explored by the first edition of Minsky & Papert (1988).

They shown that the coefficients needed to achieve linear separation (with fixed δ) sould grow very rapidly with the size of the problem and the finite number of steps needed by perceptron rule could become very large. There is after all another rule which will terminate in finite number of steps: try all integer-valued **a** in order of increasing length, and no one would advocate that rule.

Minsky & Papert also considered the behaviour of the rule when the two groups were not linearly separable, and stated that $\|\mathbf{a}\|$ would remain bounded. Thus if the **a** belong to a fixed-precision sets, the rule will eventually cycle. In particular there is no immediate way to deduce whether the rule will ever terminate, and cycling can be hard to detect, as the cycle length is unknown.

There are a number of variants of the perceptron updating rule. For example, η can be chosen just large enough to correctly classify the current case. Ho & Kashyap (1965) have other algorithms, discussed in detail in Duda & Hart (1973). It is also possible to extend the procedure to K>2 categories. In that case the natural classifier would be to choose the largest of K linear discriminants $\mathbf{x}\mathbf{a}_k$. Let **a** be the concatenation of the vectors $\mathbf{a}_k$. Then correct classification of pattern **x** in class k is equivalent to $(0\ldots,-\mathbf{x},0\ldots0,\mathbf{x}\ldots0)\mathbf{a} > 0$ with the negative element in position j, for each j not equal to k. Thus each example **x** generates g-1 examples in the K p-dimensional problem. Applying the perceptron updating rule to this problem is equivalent to the updating rule

$$\mathbf{a}_i \leftarrow \mathbf{a}_i + \mathbf{x}, \qquad \mathbf{a}_j \leftarrow \mathbf{a}_j - \mathbf{x}$$

when pattern **x** is from class i, and j is a class with a larger value of $\mathbf{x}\mathbf{a}_j$. Since this is the perceptron rule in the transformed problem, the convergence proof still holds.

4.2. Basis Function Model for Parametric Smoothing

We discuss some of the possible ways to describe more general functions of the feature variables. We consider methods using univariate functions $f : \Im \to \Re$. An additive model is of the form

$$f(\mathbf{x}) = \alpha + \sum_{j=1}^{p} g_j(x_j) \tag{4.3}$$

for smooth but unknown functions g_j (Friedman, 1989), (Hastie and Tibshirani 1990), which could encompass the effect of transformations of each feature. One choice of the smooth function $g(x)$ of a single feature is to use splines. Splines are defined by M knots ξ_i which we can consider in increasing order. Then within an interval $[\xi_i, \xi_{i+1}]$ a spline is a polynomial of degree d (often three) and at the knots the first $(d-1)$ derivatives are continuous. This can be written as

$$g(x) = \sum_{i=0}^{d} \alpha_i x^{i-1} + \sum_{i=1}^{M} \beta_i [x - \xi_i]_+^d \tag{4.4}$$

which shows that there are $M + d + 1$ free parameters. There are other bases which have better numerical properties such as B-splines (Boor, 1978). In any basis we can write

$$g(x) = \sum_{i=1}^{M+d+1} \beta_i \phi_i(x) \tag{4.5}$$

It remains to choose the parameters β_i. For a regression spline these are chosen by least squares. Cubic smoothing splines are the solution to the minimization problem

$$\sum_{i=1}^{M} [y_i - g(\xi_i)]^2 + \lambda \int g''(u)^2 \, du \tag{4.6}$$

and the parameters in (4.5) can be found by solving a sparse system of linear equations.

Additive models do not allow interactions between the features in $\Im$. Perhaps the simplest way to allow interactions is through linear combinations (projections) of features:

$$f(\mathbf{x}) = \alpha + \sum_{j=1}^{r} g_j(\alpha_j + \beta_j^T \mathbf{x}) \tag{4.7}$$

which is projection pursuit regression (PPR) (Friedman and Tukey, 1981). Sometimes the components of (4.7) are called ridge functions because a peaked g_j gives a topographic ridge in two dimensions. This is a surprisingly general class of functions, as it can approximate uniformly arbitrary continuous functions over compact. As PPR encompasses feed-forward neural networks, the functions g_j are restricted to one function, the logistic. However, ridge functions provide better approximations to some functions than others (Donoho and Johnstone, 1989), (Zhao & Atkeson, 1992), which express as working better for 'angular smooth functions' than for 'Laplacian smooth functions'. With multivariate regression we have to decide whether to use common non-linear terms for the different independent variables. This is usually done, so that for example for projection pursuit regression we have

$$f_k(\mathbf{x}) = \eta_k + \sum_{j=1}^{r} \gamma_{kj} \phi_j \left(\alpha_j + \beta_j^T \mathbf{x} \right) \tag{4.8}$$

This shows that the fitted values lie in a (r+1)-dimensional space. Since the scale of ϕ_j is not otherwise fixed, we can choose $\phi_j\left(\alpha_j + \beta_j^T \mathbf{x}\right)$ to have zero mean and unit variance over the training set.

4.3. Radial Basis Function Network

The Radial Basis Function (RBF) networks correspond to a particular class of function approximators which can be trained, using a set of samples. RBF networks have been receiving a growing amount of attention since their initial proposal (Broomhead & Lowe, 1988), (Moody & Darken, 1988), and now great deal of theoretical and empirical results are available.

4.3.1. RBF Networks Architecture

The approximation strategy used in RBF networks consists of approximating an unknown function with a linear combination of non-linear functions, called basis functions. The basis functions are radial functions, i.e., they have radial symmetry with respect to a centre. Fig. 4.1 shows a typical structure of RBF network. Let X be a vectorial space, representing the domain of the function $f(\bar{x})$ to approximate, and $\bar{x}$ a point in X. The general form for an RBF network $\aleph$ is given by the following expression:

$$\aleph(\bar{x}) = \sum_{i=1}^{n} w_i g\left(\left\|\bar{x} - \bar{c}_i\right\|_i\right) \tag{4.9}$$

where $g(z)$ is a non-linear radial function with centre in $\bar{c}_i$ and $\left\|\bar{x} - \bar{c}_i\right\|_i$ denotes the distance of $\bar{x}$ from the centre and w_i are weights. Each basis function is radial because its dependence on $\bar{x}$ is only through the term $\left\|\bar{x} - \bar{c}_i\right\|_i$.

Many alternative choices are possible for the function $g(z)$, for example, triangular, car-box or Gaussian. Anyhow it is usual to choose $g(z)$ in such a way that the following conditions hold:

$$g(-z) = g(z) \tag{4.10}$$

$$\lim_{z \to \pm\infty} g(z) = 0 \tag{4.11}$$

A common choice for the distance function $\|\cdot\|_i$ is a biquadratic form:

$$\|\bar{x}\|_i = \bar{x} Q_i \bar{x}^T \tag{4.12}$$

where Q_i is a positive definite matrix, often constrained to be diagonal:

$$Q_i = \begin{bmatrix} q_{i,11} & 0 & \cdots & 0 \\ 0 & q_{i,22} & \cdots & 0 \\ \vdots & \cdots & \cdots & \vdots \\ 0 & 0 & \cdots & q_{i,nn} \end{bmatrix}$$

In the simplest case all diagonal elements of Q_i are equal $q_{i,jj} = q_i$ so that $Q_i = q_i \mathbf{I}$. In this case the radiality of the basis functions is proper and if function $g(z)$ fades to infinity, $\frac{1}{q_i}$ can be interpreted as the width of the i-th basis function.

From the point of view of the notation is also common to write:

$$g\left(\left\|\bar{x}-\bar{c}_i\right\|_i\right)=g_i\left(\left\|\bar{x}-\bar{c}_i\right\|\right) \tag{4.13}$$

where the information about the distance function $\|\cdot\|_i$ is contained in the function $g_i(\bar{x})$.

It is also possible to define a normalized version of the RBF network:

$$\aleph(\bar{x})=\frac{\sum_{i=1}^{n} w_i g\left(\left\|\bar{x}-\bar{c}_i\right\|_i\right)}{\sum_{i=1}^{n} g\left(\left\|\bar{x}-\bar{c}_i\right\|_i\right)} \tag{4.14}$$

Different type of output, continuous or Boolean, may be needed depending on the type of the target function. In order to obtain a Boolean output $\aleph_B$ we need to compose function $\aleph$ and a derivable threshold function σ:

$$\aleph_B(\bar{x})=\sigma(\aleph(\bar{x})) \tag{4.15}$$

usually $\sigma(x)$ is the sigmoid (logistic function):

$$\sigma(x)=\frac{1}{1+e^{-kx}} \tag{4.16}$$

whose derivative is:

$$\frac{d\sigma(x)}{dx}=\sigma(x)(1-\sigma(x)) \tag{4.17}$$

The positive constant k expresses the steepness of the threshold. Alternatively, we can obtain a Boolean output composing $\aleph$ with the function $sign(x+b)$ where $b \in \Re$ is a threshold.

4.3.2. Universal Approximation

A relevant property usually required for a class of approximators is universal approximation. Given a family of function approximators, it is important to characterize the class of functions which can be effectively approximated. In general, an approximator is said to be universal if can asymptotically approximate any integrable function to a desired degree of precision.

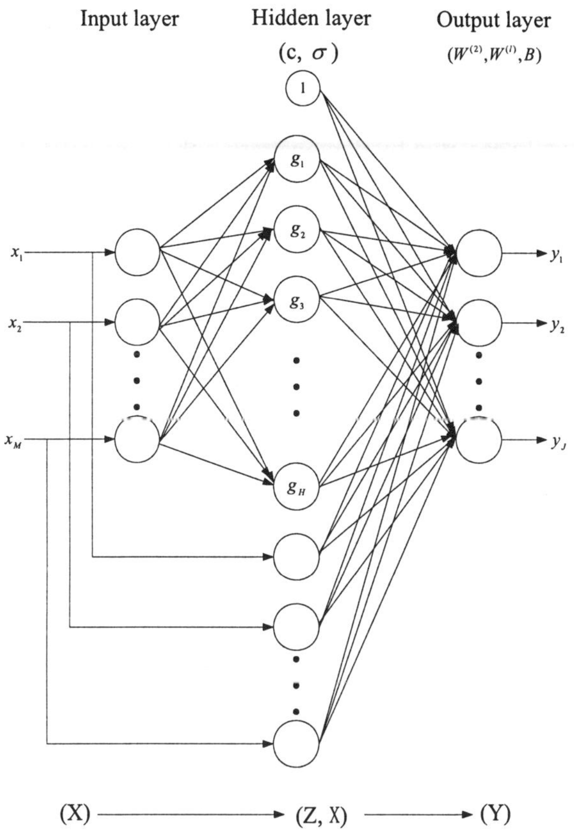

Figure 4.1. Structure of RBF network

Hornik et al. (1989) proved that any network with at least three layers, i.e., input, hidden, and output layers, is an universal approximator provided that the activation function of the hidden layer is nonlinear. In the MLP network, traditionally trained by means of the backpropagation algorithm, the most frequently used activation function is the sigmoid. RBF networks are similar to MLPs from the point of view of the topological structure but they adopt activation functions having axial symmetry.

Universal approximation capability for RBF networks was presented in (Park & Sandberg, 1993), where the problem of characterizing the kinds of radial function that entail the property of universal approximation was addressed by (Chen & Chen, 1995) who shown that for a continuous function $g(z)$ the necessary and sufficient condition is that it is not an even polynomial.

From the mathematical point of view the universal approximation property is usually asserted by demonstrating the density of the family of approximators into the set of the target functions. This guarantees the existence of an approximator that with a high, but finite number of units, can achieve an approximation with every degree of precision. The result states only that this approximator exists. It does not, however, suggest any direct method for constructing it. In general this assertion is true, even when the function is explicitly given. In other words, it is not always possible to find the best approximation within a specified class of approximators, even when the analytical expression of the function is given.

Whether the target function is Boolean or continuous, the learning task of an RBF network can be stated as a classification or regression problem. In both cases the problem can be stated in the general framework of the function approximation problem, formally expressed as: given an unknown target function $f : \Re^n \to D$ and a set S of samples (x_i, y_i) such that $f(x_i) = y_i$ for $i = 1 \ldots N$, find an approximator $\hat{f}$ of f that minimizes a cost function $E(f, \hat{f})$. Function f is a mapping from a continuous multidimensional domain X to a codomain $D \subset \Re$ (regression) or $D = B = \{0,1\}$ (classification).

The approximation accuracy is measured by the cost function $E(f, \hat{f})$ also said error function (or approximation criterion) and which depends on the set of examples *S*. In general the solution depends upon *S*, upon the choice of the approximation $\hat{f}$ is searched. In practice, a common choice for the cost function is the empirical square error:

$$E = \sum_{i=1}^{N} \left(y_i - \hat{f}(x_i)\right)^2 \tag{4.18}$$

Under some restrictive hypothesis it can be shown that minimizing (4.18) is equivalent to finding the approximator that maximizes the likelihood of S, i.e. the probability of observing S given the *a prior* hypothesis $f = \hat{f}\left(P\left\{S/f = \hat{f}\right\}\right)$ (Mitchell, 1997).

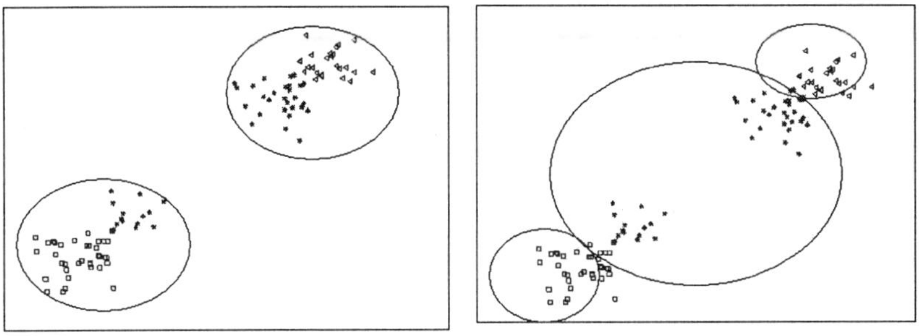

(a) (b)

Figure 4.2. Clustering of data patterns

4.3.3. Initialization and Clustering

It is critical and cumbersome to select hidden units during the initialization process. There are different approaches to tackle this problem by (Kohonen, 1997), (Kubat, 1998). Many researches suggest that the geometrical distribution of training patterns and the chosen basis function have significant effect on the number of hidden neurons (Yam & Chow, 2001). As illustrated in Fig. 4.2(a), with the unsupervised clustering algorithms (k-means, SOM), the data patterns are simply partitioned into two clusters. In the application to classification problem, it is expected that each cluster is homogeneous. The clustering in Fig. 4.2(b) is the more suitable for classification. In other words, it is expected that patterns belonging to the same class should be grouped together, while patterns from different classes should be separated. It is worth noting that unsupervised clustering algorithms are not preferable to initialize the general RBF classifier especially when one is dealing with complex data distribution (Pedrya, 1998). Hence, a supervised clustering algorithm is designed to deal with this problem.

The procedures of supervised clustering is the followings.

1. Initializing:

For every class k $(1 \le k \le m)$

Creating a new cluster S_k

Number of pattern $n_k = 1$

$$\text{Center } c_k = \frac{1}{m_k} \sum_{x_i \in \text{class k}} x_i$$

Label $l_k = k$

End For

2. Clustering:

While (the number of the left training patterns > 1)

Randomly selecting data point x from the left training patterns.

Finding out the center c_i, $\|x - c_i\| < \|x - c_j\|$, $1 \le j \le m$ and $j \ne i$

If the class of $x = i$ then

$x \in S_i$

Number of pattern $n_i = n_i + 1$

$$\text{Center } c_i = \frac{1}{n} \sum_{x_p \in S_i} x_p$$

Else

$m = m + 1$

New cluster S_{m+1};

Number of pattern $n_{m+1} = 1$;

Center $c_{m+1} = x$;

Label $l_{m+1} =$ class of x

End If

Deleting x from training patterns

End while

3. Output result

Number of cluster $N_{cluster} = m$

Data patterns labeled with $l_i(i = 1,2,\ldots, N_{cluster})$

With the above algorithm, it is likely that certain established clusters may not be able to represent real clusters in the input space. The criteria are set for eliminating the noise clusters. If the cluster S_j contains few training patterns $(n_j < ND_0)$ or occupies small space $(\sigma_j < \sigma_0)$, it is considered as 'noise' and should be deleted.

The algorithm for deleting noise clusters is as follows:

For every $S_j\left(1 \le j \le N_{cluster}\right)$

$\sigma_{j0} = \max\{\text{distant between any two data points in this cluster}\}$

If $n_j < ND_0$ Or $\sigma_{j0} < \sigma_0$

Delete S_j

Else

$\sigma_i = d_0 \times \left[\sigma_{i0}, \sigma_{i0}, \ldots, \sigma_{i0}\right]$

End If

End For

The constant d_0 affects the overlapping between different clusters. In order to avoid a too large or too small value of width for the RBF network, $d_0 = 1$ is used. In this supervised clustering algorithm, other two parameters, σ_0 and ND_0, determine the number of final clusters. The small values of them bring up a large size of clusters and will have an effect of overfitting to training data, while the relative large values of them will lead to an inaccurate description of data distribution and will degrade the classification performance. These two parameters can be determined based on the information about distribution of data patterns. σ_0 and ND_0 are determined using

$$ND_0 = \frac{\text{the number of data patterns}}{N_{cluster}}, \tag{4.19}$$

$$\sigma_0 = ND_0 \times \min\{\text{distance between any two data patterns}\}. \tag{4.20}$$

Linear parameters $W^{(1)}$, $W^{(2)}$ and B can be adjusted by the linear least squares (LLS) method as shown below:

$$\begin{bmatrix} W^{(2)} \\ B \end{bmatrix} = \begin{bmatrix} Z \\ 1 \end{bmatrix}^+ T \tag{4.21}$$

$$W^{(1)} = X^+ T \tag{4.22}$$

where $[\cdot]^+$ is defined as the pseudoinverse of $[\cdot]$, Z, X, and T are output matrix of hidden layer, network input matrix and the output target matrix respectively.

4.3.4. Learning Algorithms

The universal approximation property states that an optimal solution to the approximation problem exists: finding the corresponding minimum of the cost function is the goal of the learning algorithms. There are different learning strategies that we can follow in the design of an RBF network, depending on how the centers of the RBF of the network are specified. In the following we may assume that the choice of the RBF $g(z)$ has already been made. Essentially, we may identify three approaches, as discussed below.

4.3.4.1. Linear Weights Optimization

The simplest approach is to assume fixed RBF defining the activation functions of the hidden units. Specifically, the locations of the centers may be chosen randomly from the training data set. This is considered to be a "sensible" approach, provided that the training data are distributed in a representative manner for the problem at hand (Lowe, 1989). For the RBF themselves, we may employ an isotropic Gaussian function whose standard deviation is fixed according to the spread of the centers. A RBF centered at c_i is defined as

$$g\left(\left\|x_i - c_i\right\|^2\right) = \exp\left(-\frac{M}{d^2}\left\|x_i - c_i\right\|^2\right), \qquad i = 1,2,\ldots,M \qquad (4.23)$$

where n is the number of centers and d is the maximum distance between the chosen centers. In effect, the standard deviation (i.e., width) of all the Gaussian radial basis functions is fixed at

$$\sigma = \frac{d}{\sqrt{2M}} \qquad (4.24)$$

Such a choice for the standard deviation σ merely ensures that the Gaussian functions are not too peaked or too flat; both of these extremes are to be avoided.

The only parameters that would need to be learned in this approach are the linear weights in the output layer of the network. A straightforward procedure for doing this is to use the pseudoinverse method as

$$\mathbf{w} = \mathbf{G}^{+}\mathbf{d} \tag{4.25}$$

where **d** is the desired response vector in the training set. The matrix $\mathbf{G}^{+}$ is the pseudoinverse of the matrix **G**, which is defined as

$$\mathbf{G} = \{g_{ji}\} \tag{4.26}$$

where

$$g_{ji} = \exp\left(-\frac{M}{d^2}\left\|\mathbf{x}_j - \mathbf{c}_i\right\|^2\right), \quad j = 1,\ldots,N;\ i = 1,\ldots,M \tag{4.27}$$

where $\mathbf{x}_j$ is the jth input vector of the training set.

Basic to all algorithms for the computation of a pseudoinverse of a matrix is the singular-value decomposition:

If **G** is a real N-by-M matrix, then there exist orthogonal matrices

$$\mathbf{U} = \{\mathbf{u}_1,\ldots,\mathbf{u}_N\}$$

and

$$\mathbf{V} = \{\mathbf{v}_1,\ldots,\mathbf{v}_M\}$$

such that

$$\mathbf{U}^T\mathbf{G}\mathbf{V} = diag(\sigma_1,\ldots,\sigma_k)$$

$$k = \min(M,N) \tag{4.28}$$

The column vectors of the matrix **U** are called the left singular vectors of **G**, and the column vectors of the matrix **V** are called its right singular vectors. The $\sigma_1, \sigma_2, \ldots, \sigma_k$ are called the singular values of the matrix **G**. According to the singular value decomposition theorem, the m-by-n pseudoinverse of matrix **G** is defined by

$$\mathbf{G}^{+} = \mathbf{V}\Sigma^{+}\mathbf{U}^T \tag{4.29}$$

where Σ^{+} is an M-by-M matrix defined in terms of the singular values of **G** by

$$\Sigma^{+} = diag\left(\frac{1}{\sigma_1}, \frac{1}{\sigma_2}, \cdots \frac{1}{\sigma_k}, 0, \cdots, 0\right) \tag{4.30}$$

4.3.4.2. Gradient Descent Optimization

In this approach, the centers of the RBF and all other free parameters of the network undergo a supervised learning process, called gradient descent optimization. The first step in the development of such a learning procedure is to define the instantaneous value of the cost function

$$E = \frac{1}{2}\sum_{j=1}^{N} e_j^2 \tag{4.31}$$

where N is the number of training examples used to undertake the learning process, and e_j is the error signal, defined by

$$\begin{aligned} e_j &= d_j - \aleph(\mathbf{x}_j) \\ &= d_j - \sum_{i=1}^{n} w_i g\left(\left\|\mathbf{x}_j - \mathbf{c}_i\right\|_i\right) \end{aligned} \tag{4.32}$$

The requirement is to find the free parameters w_i, $\mathbf{c}_i$, and Σ_i^{-1} so as to minimize E. The results of this minimization are shown as below:
Linear weights in output layer

$$\frac{\partial E(n)}{\partial w_i(n)} = \sum_{j=1}^{N} e_j(n) g\left(\left\|\mathbf{x}_j - \mathbf{c}_i(n)\right\|_i\right) \tag{4.33}$$

$$w_i(n+1) = w_i(n) + \eta_w \frac{\partial E(n)}{\partial w_i(n)}, \qquad i = 1,2,\ldots,M \tag{4.34}$$

Positions of centers in hidden layer

$$\frac{\partial E(n)}{\partial \mathbf{c}_i(n)} = 2w_i(n)\sum_{j=1}^{N} e_j(n) g'\left(\left\|\mathbf{x}_j - \mathbf{c}_i(n)\right\|\right)\Sigma_i^{-1}\left[\mathbf{x}_j - \mathbf{c}_i(n)\right] \tag{4.35}$$

$$\mathbf{c}_i(n+1) = \mathbf{c}_i(n) + \eta_{\mathbf{c}} \frac{\partial E(n)}{\partial \mathbf{c}_i(n)}, \qquad i = 1,2,\ldots,M \tag{4.36}$$

Spreads of centers in hidden layer

$$\frac{\partial E(n)}{\partial \Sigma_i^{-1}(n)} = -w_i(n)\sum_{j=1}^{N} e_j(n) g'\left(\left\|\mathbf{x}_j - \mathbf{c}_i(n)\right\|\right)\mathbf{Q}_{ji}(n) \tag{4.37}$$

$$\mathbf{Q}_{ji}(n) = \left[\mathbf{x}_j - \mathbf{c}_i(n)\right]\left[\mathbf{x}_j - \mathbf{c}_i(n)\right]^T \tag{4.38}$$

$$\Sigma_i^{-1}(n+1) = \Sigma_i^{-1}(n) - \eta_\Sigma \frac{\partial E(n)}{\partial \Sigma_i^{-1}(n)} \tag{4.39}$$

The term $e_j(n)$ is the error signal of output unit j at time n. The term $g'(\cdot)$ is the first derivative of the Green's function $g(\cdot)$ with respect to its argument.

The cost function E is convex with respect to the linear parameters w_i, but nonconvex with respect to the centers $\mathbf{c}_i$ and matrix Σ_i^{-1}; in the latter case, the search for the optimum values of $\mathbf{c}_i$ and Σ_i^{-1} may get stuck at a local minimum in parameter space. The update equations shown in (4.33)-(4.39) for w_i, $\mathbf{c}_i$ and Σ_i^{-1} are assigned different learning rate parameters η_w, $\eta_\mathbf{c}$ and η_Σ, respectively. Unlike the backpropagation algorithm, the gradient descent procedure for an RBF network does not involve error backpropagation. The gradient vector $\partial E/\partial \mathbf{c}_i$ has an effect similar to a clustering effect that is task-dependent (Poggio and Girosi, 1990a).

For the initialization of the gradient descent procedure, it is often desirable to begin the search in parameter space from a structured initial condition that limits the region of parameter space to be searched to an already known useful area, which may be achieved by implementing a standard pattern classification method as an RBF network. In so doing, the likelihood of converging to an undesirable local minimum in weight space is reduced. For example, we may begin with a standard Gaussian classifier, which assumes that each pattern in each class is drawn from a full Gaussian distribution.

4.3.4.3. Hybrid of Least Squares and Penalized Optimization

For the supervising learning strategies, the RBF network optimization is usually based on a gradient descent learning process that is generally suffered from the local minima problem and slow convergence speed. In this section, a modified learning scheme is also suggested for computing the RBF parameters. This RBF learning scheme is modified from the method stated in Chapter 2 of this book (see Section 2.5.5). As stated in

the previous section, the cost function of this problem is convex with respect to the linear parameters w_i, the optimum values of this parameters are determined by the linear least squares method directly so that the convergence of the algorithm is speeded up. On the other hand, because the cost function is nonconvex with respect to the centers $\mathbf{c}_i$ and matrix Σ_i^{-1}, the optimum values of these parameters are optimized by the penalty approach so that the problem of local minima can be avoided. Based on these ideas, the RBF parameters are updated as:
For linear parameters

$$w_i(n) = \left(\Phi_j \Phi_j^T\right)^{-1}\left(\mathbf{d}_i \Phi_j^T\right) \tag{4.40}$$

where the vector

$$\Phi_j = \begin{pmatrix} 1 \\ g\left(\left\|\mathbf{x}_j - \mathbf{c}_1\right\|\right) \\ \vdots \\ g\left(\left\|\mathbf{x}_j - \mathbf{c}_M\right\|\right) \end{pmatrix}^T, \quad j = 1,2,\ldots,N, \text{ is denoted.}$$

For RBF centers

$$\mathbf{c}_i(n+1) = \mathbf{c}_i(n) + \eta_{\mathbf{c}}\left(-\frac{\partial E(n)}{\partial \mathbf{c}_i(n)} + \nabla E_{pen}\left(\mathbf{c}_i(n)\right)\right) \tag{4.41}$$

For RBF matrices

$$\Sigma_i^{-1}(n+1) = \Sigma_i^{-1}(n) + \eta_{\Sigma}\left(-\frac{\partial E(n)}{\partial \Sigma_i^{-1}(n)} + \nabla E_{pen}\left(\Sigma_i^{-1}(n)\right)\right) \tag{4.42}$$

The $\eta_{\mathbf{c}}$ and η_{Σ} are the learning rates of the RBF centers and matrices respectively. $\partial E/\partial \mathbf{c}_i$ and $\partial E/\partial \Sigma_i^{-1}$ are the error gradients with respect to $\mathbf{c}_i$ and Σ_i^{-1} respectively. $\nabla E_{pen}(\cdot)$ defines as a penalty term used for providing an uphill force under the searching space to avoid the learning process stuck in local minima. The penalty term is assigned as different types of functions as shown in Chapter 2 and the learning mechanisms are described in Section 2.5.5.

4.3.5. Regularization Networks

There is fundamental theory for RBF network in (Poggio & Girosi, 1968) which provided an elegant connection with Kolmogorov regularization theory. The basic idea of regularization consists of reducing an approximation problem to the minimization of a functional. The functional contains prior information about the nature of the unknown function, like constraints on its smoothness. The structure of the approximator is not initially given, so in the regularization framework the function approximation problem is stated as:
Find the function $F(x)$ that minimizes:

$$E(F)=\frac{1}{2}\sum_{i=1}^{n}(d_i-F(x_i))^2+\lambda\|PF\|^2=E_S(F)+\lambda E_R(F) \quad (4.43)$$

where $E_S(F)$ is the standard error term, $E_R(F)$ is the regularization term, λ is a regularization parameter and P is a differential operator.
By differentiating equation (4.43) we obtain

$$P^*PF(x)=\frac{1}{\lambda}\sum_{i=1}^{n}(d_i-F(x_i))\delta(x-x_i) \quad (4.44)$$

where $\delta(\cdot)$ is Dirac's function. The solution F of equation (4.44) is:

$$F(x)=\frac{1}{\lambda}\sum_{i=1}^{n}(d_i-F(x_i))G(x,x_i) \quad (4.45)$$

Regularization theory leads to an approximator that is an expansion on a set of Green's functions $G(x,x_i)$ of the operator P^*P. By definition Green's function of the operator A centered in x_i is

$$AG(x,x_i)=\delta(x-x_i) \quad (4.46)$$

The shape of these functions depends only on the differential operator P, i.e. on the former assumptions about the characteristics of the mapping between input and output space. Thus the choice of P completely determines the basis functions of the approximator. In particular if P is invariant for rotation and translation Green's function is:

$$G(x,x_i)=G(\|x-x_i\|) \quad (4.47)$$

so they depend only on the distance $\|x - x_i\|$ and are therefore radial functions.

The point x_i are the centers of the expansion and the term $\frac{1}{\lambda}(d_i - F(x_i))$ of equation (4.47) are the coefficients.

The approximator is

$$w_i = \frac{1}{\lambda}(d_i - F(x_i)) \qquad F(x) = \sum_{i=1}^{n} w_i G(x, x_i) \tag{4.48}$$

Equation (4.48) evaluated in the point x_j leads to

$$F(x_j) = \sum_{i=1}^{n} w_i G(x_j, x_i) \tag{4.49}$$

In order to determine the w_i let us define the matrices:

$$F = \begin{bmatrix} F(x_1) \\ F(x_2) \\ \vdots \\ F(x_N) \end{bmatrix} \qquad d = \begin{bmatrix} d_1 \\ d_2 \\ \vdots \\ d_N \end{bmatrix}$$

$$W = \begin{bmatrix} w_1 \\ w_2 \\ \vdots \\ w_N \end{bmatrix} \qquad G = \begin{bmatrix} G(x_1, x_1) & G(x_1, x_2) & \cdots & G(x_1, x_N) \\ G(x_2, x_1) & G(x_2, x_2) & \cdots & G(x_2, x_N) \\ \vdots & \vdots & \ddots & \vdots \\ G(x_N, x_1) & G(x_N, x_2) & \cdots & G(x_N, x_N) \end{bmatrix} \tag{4.50}$$

Then equation (4.49) can be represented in the form of matrices:

$$W = \frac{1}{\lambda}(d - F) \tag{4.51}$$

$$F = GW \tag{4.52}$$

Eliminating **F** from both expressions, we obtain:

$$(G + \lambda I)W = d \tag{4.53}$$

The matrix **G** is symmetric and for some operator is positive definite. It is always possible to choose a proper value of λ such that $\mathbf{G}+\lambda\mathbf{I}$ is invertible, that leads to:

$$\mathbf{W}=(\mathbf{G}+\lambda\mathbf{I})^{-1}\mathbf{d} \tag{4.54}$$

It is not necessary to expand the approximator over the whole data set, in fact the point x_i on which equation (4.48) was evaluated is arbitrarily chosen. If we consider a more general case in which the centers of the basis functions c_i with $i=1\ldots N$ are distinct from the data the matrix **G** is rectangular. Defining two new matrices as:

$$\mathbf{G}_0=\left(G\left(c_i,c_j\right)\right) \qquad i,j=1\ldots N \tag{4.55}$$

$$\mathbf{G}=\left(G\left(x_i,c_j\right)\right) \qquad i=1\ldots M \quad j=1\ldots N \tag{4.56}$$

The optimal weights are:

$$w=\left(\mathbf{G}^T\mathbf{G}+\lambda\mathbf{G}_0\right)^{-1}\mathbf{G}^T\mathbf{d} \tag{4.57}$$

and if $\lambda=0$

$$w=\left(\mathbf{G}^T\mathbf{G}\right)^{-1}\mathbf{G}^T\mathbf{d}=\mathbf{G}^{+}\mathbf{d} \tag{4.58}$$

where $\mathbf{G}^{+}=\left(\mathbf{G}^T\mathbf{G}\right)^{-1}\mathbf{G}^T\mathbf{d}$ is the pseudoinverse matrix.
In the regularization framework the choice of the differential operator P determines the shape of the basis function. Haykin (1994) reports a formal description of the operator that leads to the Gaussian RBF network. The operator expresses conditions on the absolute value of the derivatives of the approximator. Hence the minimization of the regularization term $E_R(F)$ causes a smoothing of the function encoded by the approximator.

4.4. Advanced Radial Basis Function Networks

4.4.1. Support Vector Machine

Radial Basis Function (RBF) networks are deeply related to Supper Vector Machines (SVM) those are learners based Statistical Learning

Theory (Vapnik, 1995). In the case of classification the decision surface of a SVM is given in general by

$$SVM(\bar{x}) = sign(\bar{w}\,\phi(\bar{x}) + b) \tag{4.59}$$

where $\phi : \Re^n \rightarrow F$ is a mapping in some feature space F. The parameters $\bar{w} \in F$ and $b \in R$ are such that they minimize an upper bound on the expected risk. We omit the formula of the bound that represents a fundamental contribute given by Vapnik to statistical learning theory. For the present purpose it suffices to remember that the bound is composed by an empirical risk term and a complexity term that depends on the VC dimension of the linear separator. Controlling or minimizing both the terms permits control over the generalization error in a theoretically well-founded way.

The learning procedure of a SVM can be sketched as follows. The minimization of complexity term can be achieved by minimizing the quantity $\frac{1}{2}\|\bar{w}\|^2$, namely the square of the norm of the vector $\bar{w}$. In addition the strategy is to control the empirical risk term by constraining:

$$y_i(\bar{w}\,\phi(\bar{x}_i) + b) \geq 1 - \mu_i \tag{4.60}$$

with $\mu_i \geq 0$ and $i = 1 \ldots N$ for each sample of the training set. The presence of the variables μ_i allows some misclassification on the training set.

Introducing a set of Lagrange multipliers α_i, $i = 1 \ldots N$ if it is possible to solve the programming problem defined above, finding $\bar{w}$, the multipliers and the threshold term b. The vector $\bar{w}$ has the form:

$$\bar{w} = \sum_{i=1}^{N} \alpha_i y_i \phi(\bar{x}_i) \tag{4.61}$$

so the decision surface is:

$$SVM(\bar{x}) = sign\left(\sum_{i=1}^{N} \alpha_i y_i \phi(\bar{x}_i)\phi(\bar{x}) + b\right) \tag{4.62}$$

where the mapping ϕ compares only in the dot product $\phi(\bar{x}_i)\phi(\bar{x})$. The dependency only on the dot product and not on the mapping ϕ is valid also for the multipliers. Following (Mueller et al., 2001), the connection between RBF networks and SVMs is based on upon the remark that a kernel function $k(\bar{x},\bar{y})$ defined on $k: C \times C \to R$ with C a compact set of $\Re^n$, namely

$$\forall f \in L^2(C): \int_C k(\bar{x},\bar{y}) f(\bar{x}) f(\bar{y}) d\bar{x} d\bar{y} \geq 0 \qquad (4.63)$$

can be seen as the dot product $k(\bar{x},\bar{y}) = \phi(\bar{x})\phi(\bar{y})$ of a mapping $\phi: \Re^n \to F$ in some feature space F. As a consequence, it is possible to substitute $k(\bar{x},\bar{y}) = \phi(\bar{x})\phi(\bar{y})$ obtaining the decision surface expressed as:

$$SVM(\bar{x}) = sign\left(\sum_{i=1}^{N} \alpha_i y_i k(\bar{x}_i, \bar{x}) + b\right) \qquad (4.64)$$

Choosing a radial kernel $k(\bar{x},\bar{y}) = e(\|\bar{x} - \bar{y}\|)$ such that the equation has the same structure of the RBF network presented in section 4.3.1 in the case of a classification task. The possibility of interpreting RBF networks as an SVM permits application of this technique to control complexity and prevent overfitting. Complexity regularization has also been studied directly for RBF networks (Kryzak & Linder, 1999) with bounds on the expected risk in terms of the sample set size. SVMs also connect RBF networks with Kernel-Based Algorithms and following (Muller et al., 2001) with Boosting techniques.

4.4.2. Wavelet Network

Wavelet network is a network combining the ideas of the feed-forward neural networks and the wavelet decompositions, Zhang and Benveniste (1992) provide an alternative to the feedforward neural networks for approximating functions. Wavelet networks use simple wavelets and wavelet network learning is performed by the standard backpropagation type algorithm as the tradition neural network. The localization property of wavelet decomposition is reflected in the important properties of wavelet networks. Wavelet networks can approximate any continuous functions on $[0,1]^n$ and have certain advantages such as the use of wavelet coefficients as the initial value for backpropagation training and possible reduction of the network size while achieving the same level of approximation. In a feedforward network, neurons take their inputs from the previous layer only and send the outputs to the next layer only. Since the signals go in one direction only, the network can compute a result very quickly. Basic neuron of a wavelet network are multidimensional wavelets and the neuron parameters are the dilation and translation coefficients. The output of a wavelet network is the linear combination of the values of several multidimensional wavelets.

Suppose there is a function ψ define on $\Re^n$ such that there is a countable set Ψ of the form

$$\Psi = \{\psi(D_i(\mathbf{x}-\mathbf{t}_i))\} \tag{4.65}$$

D_i is an $n \times n$ diagonal matrix with the diagonal vector $d_i \in \Re^n$, $\mathbf{x}$ and $\mathbf{t}_i \in R^n$, is a frame, which means there exist constants A and B such that $A\|f\|^2 \le \sum \alpha \in \Psi|\langle \alpha, f\rangle|^2 \le B\|f\|^2$ for any $f \in L^2\Re^n$. It follows from the frame property that the set S of all linear combinations of the elements in Ψ is dense in $L^2\Re^n$. Obviously, the set of all linear combinations of the form

$$\sum_{i=1}^{N} w_i \psi(D_i(\mathbf{x}-\mathbf{t}_i)) \tag{4.66}$$

where D_i and $\mathbf{t}_i$ are not restricted to those in Ψ is a superset of S and is also dense in $L^2\Re^n$. For example, we can use ψ given by

$$\psi(\mathbf{x}) = \psi(x_1,\ldots,x_n) = \psi_S(x_1)\psi_S(x_2)\cdots\psi_S(x_n) \tag{4.67}$$

where ψ_S is the function given by

$$\psi_S(x) = -xe^{-x^2/2} \tag{4.68}$$

the first derivative of the Gaussian function $e^{-x^2/2}$. Note the derivative of ψ_S,

$$\frac{d}{dx}\psi_S(x) = -e^{-x^2/2}\left(1-x^2\right) \tag{4.69}$$

The wavelet network structure will be of the form

$$h(w,\mathbf{x}) = \sum_{i=1}^{N} a_i \psi(D_i(\mathbf{x}-\mathbf{t}_i)) + \overline{g} \tag{4.70}$$

where $a_i \in R$, ψ is a given wavelet function, D_i is an $n \times n$ diagonal matrix, $\mathbf{x}$ and $\mathbf{t}_i \in R^n$, and $\overline{g}$ the average value of $g(\mathbf{x})$. w represents all the parameters $a_1,\ldots,a_n$, $D_1,\ldots,D_n$, $t_1,\ldots,t_n$, and $\overline{g}$. The matrix D_i and $\mathbf{t}_i$ are set by the wavelet decomposition and the weights w_i are initially set to be zero. It should be noted that the wavelet decomposition uses the given D_i and $\mathbf{t}_i$ and finds the weight coefficients w_i, while the wavelet network tries to adjust D_i, $\mathbf{t}_i$ and the weight coefficients w_i altogether to fit the data. Wavelet networks can be used for concept learning for a concept $S \subset [0,1]^n$ using $f = \chi s$, the characteristic function of S, that is,

$$\begin{aligned} f(\mathbf{x}) &= 1 \quad \text{for } \mathbf{x} \in S \\ &= 0 \quad \text{for } \mathbf{x} \in S. \end{aligned} \tag{4.71}$$

Given $\varepsilon > 0$, there exist

$$g(\mathbf{x}) = h(w,\mathbf{x}) = \sum_{i=1}^{N} a_i \psi(D_i(\mathbf{x}-\mathbf{t}_i)) + \overline{g} \tag{4.72}$$

and $D \subset [0,1]^n$ with measure $\geq 1-\varepsilon$ such that

$$|g(\mathbf{x}) - f(\mathbf{x})| < \varepsilon \qquad \forall \mathbf{x} \in D \tag{4.73}$$

The learning algorithm of a wavelet network modifies the dilation and translation coefficients of every wavelet neuron and the coefficients (weights) of the linear combination of the neurons so that the network closely fits the data. We assume the data is contaminated with noise, so the learning algorithm should not seek to interpolate the data points. The network g_θ, where θ represents all the parameters D_i, $\mathbf{t}_i$ and w_i, will be adjusted by the learning algorithm to minimize a suitable objective function, so that it becomes an optimization problem. A simple objective function we consider is

$$C(\theta) = E\left(\left|g_\theta(\mathbf{x}) - y\right|^2\right) \tag{4.74}$$

where $\mathbf{x}_k$ and y_k are data pairs (that is, $f(\mathbf{x}_k) = y_k + n_k$, where n_k is a random noise). Though a standard gradient descent algorithm can be used, a heavy computation requirement makes it impractical in some situations. In practice, some other more efficient algorithms, such as stochastic gradient method, are used. The function computed by the basic wavelet network model is differentiable with respect to all parameters (dilation and translation parameter and the weights).

4.4.3. Fuzzy RBF Controllers

Radial Basis Function networks also are interpreted as fuzzy controllers (Jang, 1993). In general, a controller of this kind is a software or hardware implementation of a control function, defined from the state-space of the system to its input-space. In this way, the control function maps a set of information about the state of the system we want to control, to the actions the controller has to apply to the system. Typically, the state and the actions are continuous vectors and the controller is fully described by a set of input variables X, a set of output variables Y, and the set of elements implementing the control function. In the case of fuzzy controllers, the internal elements are defined by means of a fuzzy logic propositional theory.

Fuzzy logics are based on a generalization of the characteristic function of a set. Formally, let f_A be the characteristic function of a set A:

$$f_A(x) = \begin{cases} 1 & \text{if } x \in A \\ 0 & \text{if } x \notin A \end{cases} \tag{4.75}$$

Fuzzy set theory (Zadeh, 1965) generalizes the notion of presence of an element in a set and consequently the notion of characteristic function, by introducing fuzzy values. This approach is equivalent to introducing uncertainty in the presence of an element in a set. The fuzzy characteristic function is called membership which can assume any value in the interval [0, 1]. A set in which the membership function is restricted to assume the values in the set {0, 1}, is said to be crisp. The introduction of a fuzzy membership has deep implications concerning the logics which can be built on it. The first one is the possibility of have fuzzy truth values for predicates. A predicate is no longer simply false (0) or true (1) but can assume any value between. Consequently, the definitions of the basic connectives (disjunction, conjunction and negation) have to deal with fuzzy values. Fuzzy logics are typically used for expressing uncertain or approximate knowledge in the form of rules. The theory can be partially contradictory, causing fuzzy memberships to overlap each other. Many different shapes for the membership functions have been proposed (triangular, trapezoidal, Gaussian) [see (Berebji, 1992)].

Usually a fuzzy controller is organized as three layers. The first one implements the so-called fuzzification operation and maps every dimension of the input space via the memberships, to one or more

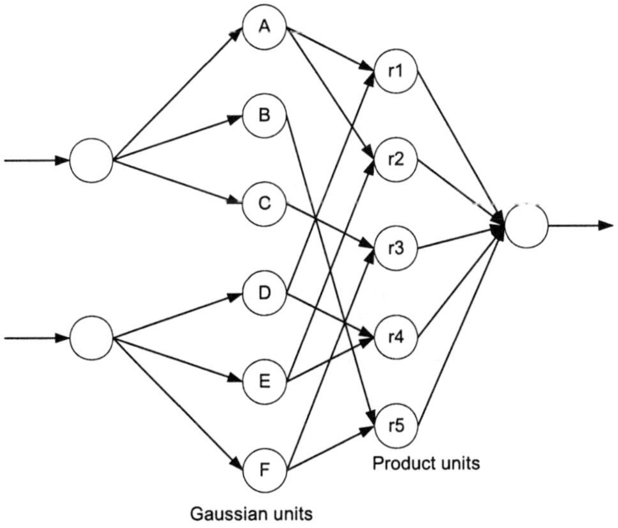

Figure 4.3. Fuzzy RBF network architecture. The first layer hidden units have a one-dimensional Gaussian activation function. The second layer hidden units compose the input values using arithmetic product. An average sum unit performs the weighted sum of the activation values received from the product units

linguistic variables, in a fuzzy logic language. The linguistic variables are then combined with the fuzzy connectives to form the fuzzy theory. Typically the theory is propositional and it implements the defuzzification transforming back the continuous truth values into points in the output space. A RBF based fuzzy controller (Blanzieri & Giordana, 1995) was introduced in which the architecture is also similar to the fuzzy neural networks for implementing fuzzy controllers capable of learning from a reinforcement signal, and to the architecture proposed by Tresp et al. (1993). Fig. 4.3 describes the basic network topology.

The activation function used in a fuzzy RBF network with n input units is defined as the product of n one-dimensional radial functions, each one associated to one of the input features. Therefore a fuzzy RBF can be described as a network with two hidden layers. The neurons in the first hidden layer are feature detectors, each associated to a single one-dimensional activation function and are connected to a single input only. For example, if we choose to use Gaussian functions, the neuron r_{ij} (the i-th component of the j-th activation area) computes the output:

$$\mu_{ij} = e^{-\left(\frac{I_i - C_{ij}}{\sigma_{ij}}\right)^2} \tag{4.76}$$

The neurons in the second hidden layer simply compute a product and construct multi-dimensional radial functions:

$$r_j = \prod_i \mu_{ij} = g_j \tag{4.77}$$

where g_j was introduced in section 4.3.1.

Finally, the output neuron combines the contributions of the composite functions computed in the second hidden layer. In this architecture, a choice of four different activation functions is possible for the output unit, in order to adapt the network to different needs. The output function can be a weighted sum

$$Y = \sum_j w_j r_j \tag{4.78}$$

The same function can be followed by a sigmoid when the network is used for a classification task. Using this function the network tends to produce an output value close to '0' everywhere the input vector falls in a point of the domain which is far from every activation area. The consequence is under-generalization in the classification tasks. This problem can be avoided by introducing a normalization term in the output activation function:

$$\hat{Y} = \frac{\sum_j w_j r_j}{\sum_j r_j} \tag{4.79}$$

This function is frequently used for fuzzy controller architectures. In this case, one obtains a network biased toward over-generalization in a similar way as for the multi-layer perceptron. Depending on the application, under-generalization or over-generalization can be preferable.

4.4.4. Probabilistic Neural Networks

Probabilistic Neural Networks (PNN) originate in a pattern recognition framework as tools for building up classifiers. In that framework the examples of a classification problem are points in a continuous space and they along to two different classes conventionally named 0 and 1. PNN were first proposed by Specht (1990), who proposed to approximate, separately, the density distributions $g_1(\bar{x})$ and $g_0(\bar{x})$ of the two classes and use a Bayes strategy for predicting the class

$$\hat{f}(\bar{x}) = \begin{cases} 1 & \text{if } p_1 l_1 g_1(\bar{x}) > p_0 l_0 g_0(\bar{x}) \\ 0 & \text{if } p_1 l_1 g_1(\bar{x}) < p_0 l_0 g_0(\bar{x}) \end{cases} \tag{4.80}$$

where p_1 and p_0 are the a priori probabilities for the classes to separate and, l_1 and l_0 are the losses associated with their misclassification (l_1 loss associated with the decision $\hat{f}(\bar{x}) = 0$ when $\hat{f}(\bar{x}) = 1$).
Then the decision surface is described by the equation:

$$g_1(\bar{x}) = k g_0(\bar{x}) \tag{4.81}$$

where $k = \dfrac{p_0 l_0}{p_1 l_1}$ and defining $\sigma(x)$ as a threshold function the estimate of the target function is:

$$\hat{f}(\bar{x}) = \sigma(g_1(\bar{x}) - kg_0(\bar{x})) \tag{4.82}$$

Again the density approximations are made using the kernel estimations

$$g_1(\bar{x}) = \frac{1}{N_1|H|}\sum_{i=1}^{N_1} K_{n+1}\left(H^{-1}(\bar{z} - \bar{z}_1)\right) \tag{4.83}$$

with the extension of the sum limited to the N_1 instances belonging to class 1 and analogously for the class 0.

$$\hat{f}(\bar{x}) = \sigma\left(\frac{1}{|H|}\sum_{i=1}^{N} C(z_i) K_{n+1}\left(H^{-1}(\bar{z} - \bar{z}_i)\right)\right) \tag{4.84}$$

where

$$C(z_i) = \begin{cases} 1 & \text{if } f(z_i) = 1 \\ -k\dfrac{N_1}{N_0} & \text{if } f(z_i) = 0 \end{cases} \tag{4.85}$$

The equation (4.80) is a particular case of the RBF network described in equation (4.14) in Section 4.3.1 for approximating Boolean functions.

In the statistical framework it is common to use all the data as centers of the kernels. In the case of a large data set it is possible to limit the initialization to an extracted sample of data. It is worth noting, that no computation is needed to find the values of the weights. In fact, as an effect of the normalization terms contained in the kernels, the weights are equal to the output values, or set to an estimate of the a priori probability of the class. This method can be applied in an incremental way, but like any other method which uses all the data, it suffers for the overgrowing of the approximator. This property permits to analyze directly the network in terms of probability having a direct statistical interpretation. In fact, given the estimation of the density implicitly formed by the RBF network is possible estimate all the statistics parameter of the distribution. Moreover exploiting a factorizable architecture is possible to express independency between the inputs as is normally done in Bayesian Networks.

4.5. Concluding Remarks

This chapter is concerned with Radial Basis Function (RBF) networks and it addresses their different interpretations and the applicative perspectives as a classifier. RBF networks can be described as three layer neural networks where hidden units have a radial activation function. Although some of the results of the neural networks can be extended to RBF network, exploiting this interpretation (e.g. approximation capabilities and the existence of a unique minimum (Bianchini et al., 1995), substantial difference still remain with respect to the other feedforward networks. In fact, RBF networks exhibit properties substantially different with respect to both learning properties and semantic interpretation. In order to understand the different behaviours of the two network types, assume we have to modify a weight between two nodes in the multilayer perceptron, as is done by the backpropagation updating rule during the training phase. The effect involves an infinite region of the input space and can affect a large part of the co-domain of the target function. On the contrary, changing the amplitude of the region of the input space in which the activation function of a neuron in an RBF network fires, or shifting its position, will have an effect local to the region dominated by that neuron. More in general, this locality property of RBF networks allows the network layout to be incrementally constructed, adjusting the existing neurons, and/or adding new ones. As every change has a local effect, the knowledge encoded in the other parts of the network is not lost; so, it will not be necessary to go through a global revision process.

An important point of the present work is the systematic way the different interpretations have been presented in order to permit their comparison. RBF networks are particularly suitable for integrating the symbolic and connectionist paradigms in the line drawn by Towell and Shavlik (1994) whose recent developments has been surveyed by Cloete and Zurada (2000). This symbolic interpretation permits to consider RBF networks as intrinsically Knowledge-Based Networks. Moreover, RBF networks have also very different interpretations. They are Regularization Networks so there is the possibility of tuning the regularization parameter. They are Support Vector Machines so they gain theoretical foundation from statistical learning theory. They are related to Wavelet Networks so they can gain advantage in signal applications. They have a Fuzzy interpretation so they can be interpreted in terms of fuzzy logic. They have a statistical interpretation so they can produce, after training, knowledge in terms of probability. They are also instance-based learners and so they can provide a case-based reasoning modality. Finally, another basic property of the RBF network is the locality that permits the synthesis of incremental dynamic algorithms permitting the growing of the cases without unlearning.

Exercises

Q4.1. Consider a 2-input, 4-hidden and 1-ouput nodes of Radial Basis Function network for XOR function. Compute the linear weight **w** and sketch the profiles of the radial basis functions if linear RBF operators are used:

$$\phi_i(\underline{x}) = \| \underline{x} - \underline{x}_i \|$$

Q4.2. Repeat the exercise of Q4.1 if Gaussian RBF operators are used:

$$\phi_i(\underline{x}) = \exp(-\frac{\| \underline{x} - \underline{x}_i \|^2}{2\sigma})$$

where σ=1.

Q4.3. Compare between the linear separation capabilities of RBF network and MLP network.

Q4.4. The gradient descent optimization can be used for optimizing the RBF parameters. In order to compute the gradients, it is required to compute the derivative g'(h) of the RBF operators g(h). Suppose a Gaussian RBF operator is used, compute the derivative of this operator.

Q4.5. For a Support Vector Machine (SVM), given a training data sample $\{(\mathbf{x}_i, \mathbf{d}_i),\ i = 1, \ldots, N\}$, find the optimum values of the weight vector **w**

$$\mathbf{w} = \sum a_{0,i} d_i \phi(\mathbf{x}_i)$$

where $a_{0,i}$ are the optimal Lagrange multipliers determined by maximising the following objective function

$$Q(a) = \sum_{i=1}^{N} a_i - \frac{1}{2}\sum_{i=1}^{N}\sum_{j=1}^{N} a_i a_j d_i d_j \underline{\phi}^T(\underline{x}_i)\underline{\phi}(\underline{x}_j)$$

subject to the constraints

$$\sum a_i d_i = 0; \quad a_i > 0$$

Let the nonlinear mapping for the XOR function be:

$$\phi(\mathbf{x}) = \left(1, x_1^{\ 2}, \sqrt{2}x_1 x_2, x_2^{\ 2}, \sqrt{2}x_1, \sqrt{2}x_2\right)^T$$

And $\phi(\mathbf{x}_i)\left(1, x_{i1}^{\ 2}, \sqrt{2}x_{i1}x_{i2}, x_{i2}^{\ 2}, \sqrt{2}x_{i1}, \sqrt{2}x_{i2}\right)^T$

The feature space for XOR function is in 6D with 20 input data

$$\mathbf{x}_1 = (-1,-1), \quad d_1 = -1$$
$$\mathbf{x}_2 = (-1,1), \quad d_2 = 1$$
$$\mathbf{x}_3 = (1,-1), \quad d_3 = 1$$
$$\mathbf{x}_4 = (1,1), \quad d_4 = -1$$

Minimize the above cost function to find the optimal decision boundary for this XOR problem.

Chapter 5

Self-organizing Maps

5.1. Introduction

The Self-Organizing Map (SOM), also known as the Kohonen feature map, was introduced by Kohonen in 1982. In contrast to many other neural networks, which usually require a teacher and a supervising learning process, it is an unsupervised network that does not require a teacher. Its outputs are organized in a way of groups or clusters. It is particular useful for clustering high dimensional data. For instance, consider a simple example of analyzing colour grouping. If an SOM with 3 input nodes connected to random Red, Green, and Blue values representing different RGB values of a color, the SOM output map ultimately contains the high level colours such as Red, Orange, Yellow, Green, Blue, Violet, etc. The Red, Green, and Blue will be clustered in corners, whilst other colors such as, Yellow, Orange, Violet will be clustered in between. The SOM can also be seen as a combination of vector quantization and dimension reduction method in one algorithm and can be used for visualization or projecting high-dimensional data to low dimensions. Fig. 5.1(a) shows color distributions among neurons of a randomly initialized 20x20 SOM by assigning random 3 RGB values to their weight vector. After the training is complete by randomly generated color data, the three basic colors red, green, blue appeared in three corners and others intermediate colors appeared in-between as shown in Fig. 5.1(b). This example shows the organizing capability, which is referred to "Topological ordering", of the SOM.

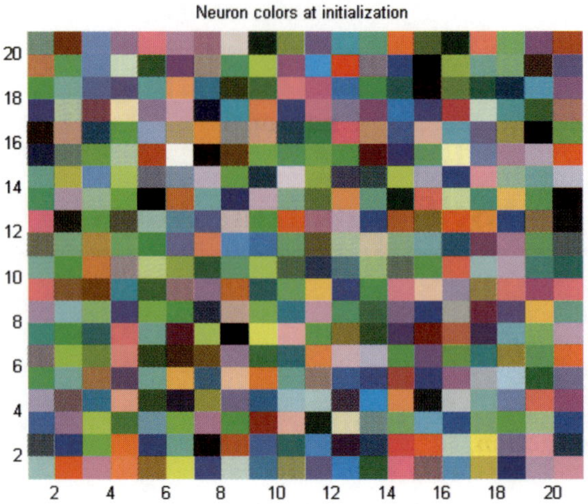

(a)

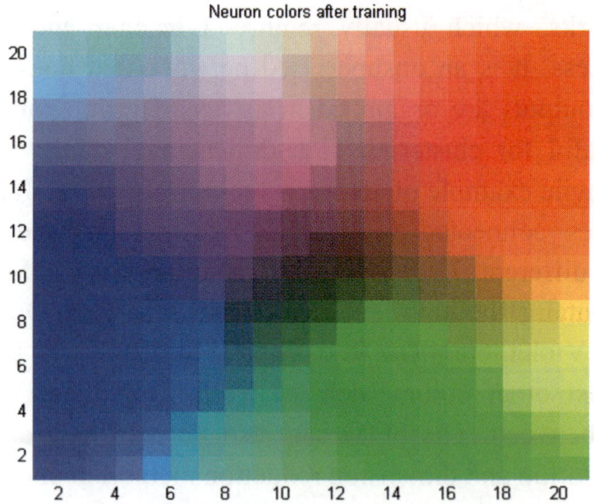

(b)

Figure 5.1. Color distribution among neurons (a) at random initialization and (b) after the training is complete

The SOM has been widely used in many areas such as pattern recognition, biological modeling, data compression, and data mining. The success of the SOM algorithms lies with its simplicity making it easy to understand. Usually, it is considered as a heuristic approach because its fundamental theories have not been derived from strict mathematics. The basic SOM algorithm consists of a set of neurons usually arranged in a 2-dimentional grid such that there are neighborhood relations among neurons. Each neuron is attached to a feature vector of the same dimension as the input space. The neurons are selectively adjusted to various inputs or input patterns during the course of a competitive training in which the neurons compete among themselves. The weight vectors of winning neurons and their neighboring neurons are then adjusted systematically. As a result, a meaningful systematic coordinate map will be established for representing the given input features. Through assigning each input data to the neuron with the nearest feature vector, an SOM is able to group input data into regions with the similar feature vectors. This process can be considered as vector quantization. Also, because of the neighborhood relation contributed by the inter-connections among neurons, it exhibits an important property of topology preservation. In other words, when feature vectors are near from each other in the input space, the corresponding neurons will also be close in the output space.

Since the SOM was first introduced by Kohonen, it has undergone many modifications. Before briefing some of these recent works, we need to describe some of the deficiencies of classical SOM. Generally we need to pre-define the map structure and the map size prior to the commencement of the training process. Conventional SOM topology seems to be inherently limited by the fixed network. One must adopt a number of trials and tests to select an appropriate network structure and size. Several improved SOMs or related algorithms have been developed to overcome these shortcomings. All these algorithms are mainly in the direction of growing an SOM adaptively. Although most of these extended algorithms are able to dynamically increase the network to an appropriate size, the final SOM maps are either not easy to visualize high-dimensional input data on a 2-D plane, or are of equal distances among neighboring neurons in a 2-D output map.

Indeed, the SOM can be seen as discrete approximation of principal surfaces in the input space. The ViSOM, a new visualization method, regularizes the inter-neuron distances such that the inter-neuron distances in the input space resemble those in the output space after the completion of training. This feature can be useful to some applications because it is able to preserve the topology information as well as the inter-neuron distances. This characteristic is attributed to the output topology pre-defined in a regular 2-D grid so that the trained neurons are almost regularly distributed in the input space. The ViSOM delivers excellent visual exhibition compared with conventional SOM and other visualization methods.

The SOM is also used for clustering data. In Huntsberger and Ajjimarangsee (1989), and Mao and Jain (1996), an SOM was used to develop clustering algorithms. There is also a two-level SOM approach used to cluster data. The first level is used to train data and the second level is used to cluster data. The number of output neurons required at the first level is more than the desired number of clusters. Clustering is carried out by clustering of output neurons after completion of training. This two-level approach has been discussed by Lampinen and Oja (1992), Kiang (2001) and several other researchers.

The SOM is widely known with its ability of performing visualization. It is useful in data mining and facilitating people in understanding data visually. Through visualization we can evaluate mined patterns and finally unearth the truly interesting or useful patterns. There have been quite a few SOM based visualization methods reported in different literatures. The assignment method simply assigns inputs to their corresponding neurons. The U-matrix utilizes inter-neuron distances to show the clustering tendency of data. There are other methods attempting to find a non-uniformly distributed 2-D output map, instead of a uniformly distributed output neurons. In later sections of this chapter, we will introduce and discuss several modified SOM topologies.

5.2. Self-Organizing Maps

The self-organizing map consists of a set of neurons usually arranged in a one or two dimensional grid. Through a competitive learning, the weights of neurons are adjusted to various input patterns while maintaining topological order in the output map. Architecture of a typical 2-dimensional SOM is shown in Fig. 5.2.

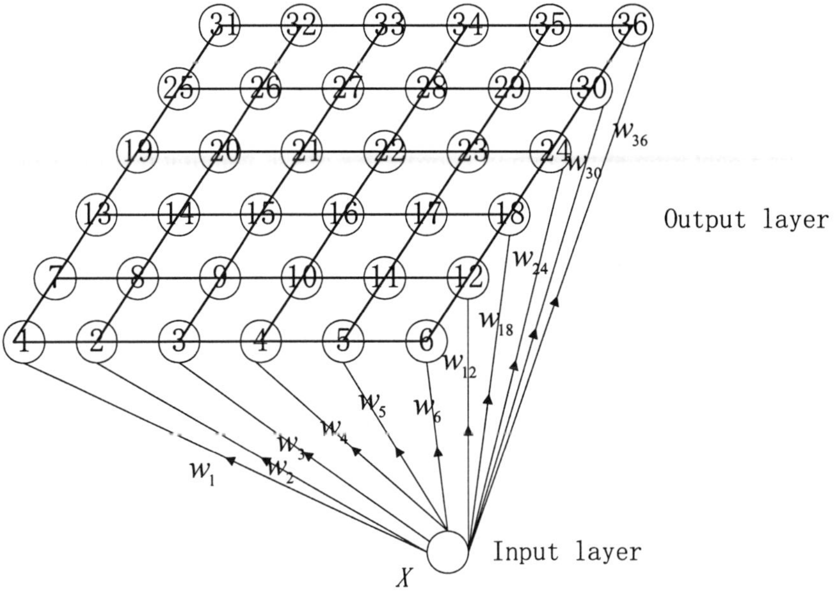

Figure 5.2. Architecture of a 2-dimensional SOM

In this configuration, input data are presented through the input layer. The neurons in the 2-dimensional output layer are fully connected with input layer. This means the synaptic weight of each neuron has the same dimension of the input data.

5.2.1. *Learning Algorithm*

For each input vector $x(t)$, the SOM algorithm first finds the closest neuron c on the output map of size N by

$$c = \arg\min_i \|x(t) - w_i\|, \; i = 1, \; 2, \; \dots, \; N$$

where w_i is the weight vector of the ith neuron.

A set of neighboring neurons of the winning neuron is denoted as N_c. The neighboring radius of the winning neuron c decreases with time. The sequential weight-updating rule of the SOM algorithm is given by

$$w_i(\mathrm{t}+1)=\begin{cases} w_i(t)+\varepsilon(t)h(c,i;t)(x(t)-w_i(t)), & \forall i \in N_c \\ w_i(t)\,, & \text{otherwise} \end{cases} \tag{5.1}$$

where $\varepsilon(t)$ is the learning rate and $h(c,i;\ t)$ is the neighborhood function centering at the output coordinates of the winning neuron. Both $\varepsilon(t)$ and $h(c,i;\ t)$ decreases with time. A typical settings for $h(c,i;\ t)$ is: $h(i,j;t)=\exp\left(-\frac{d(i,j)^2}{2\sigma^2(t)}\right)$, where t is the iteration number, $d(i,j)$ is the distance of each neuron from the winner, and $\sigma(t)$ is the width of the neighborhood function $h(c,i;\ t)$. Usually $\sigma(t)=\sigma_0.\exp\left(-\frac{t}{\tau_1}\right)$, where τ_1 is a time constant and σ_0 is the initial width. The learning rate of the training typically controlled as: $\varepsilon=\varepsilon_0.\exp\left(-\frac{n}{\tau_2}\right)$, where τ_2 is the time constant and ε_0 is the initial learning rate. The weight updating for all data repeats until the maximum number of iterations is reached.

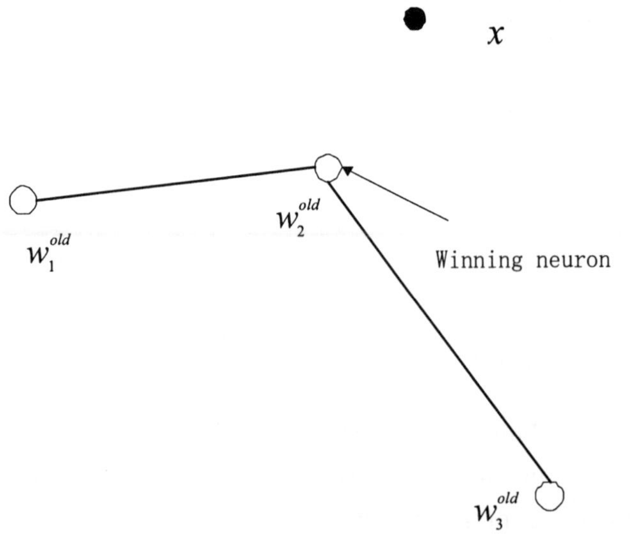

Figure 5.3. Weight distributions before weight updating

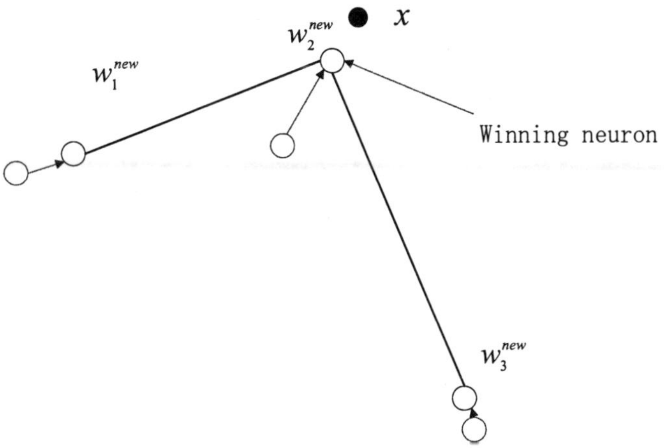

Figure 5.4. Weight distributions after weight-updating

We use an example of three neurons in the data space to illustrate the learning strategy. In Fig. 5.3, x is the current input datum. The weights of the three neurons are w_1, w_2 and w_3. Before updating, it can be noticed that w_2 is the closest neuron to input datum before updating. Neuron 2 is the winner neuron and it learns more than the other two neurons, i.e., $h_2 = 1$, $h_2 >> h_1$ and $h_2 >> h_3$, where h_1, h_2, h_3 are the neighborhood function for neurons 1, 2 and 3 respectively. As a result, the updated w_2 becomes closer to the datum x and the displacement of w_2 is bigger than that of the other two weights shown in Fig. 5.4.

Let us consider another example whose input is a uniformly distributed 2-D synthetic data in a unit square (400 data). In this example, an SOM is used to form an ordered 2-D net interpolated in a 2-D input space.

Fig. 5.5 shows the distribution of input data and initial neurons' weight as well as the neighborhood connections between neighboring neurons (Map size 8×8). After the SOM has been updated for 500 iterations, the weights distribution is shown in Fig. 5.6.

When we observe the weights distributions and their interconnection in Fig. 5.5 and Fig. 5.6, the weights become more ordered in the data space after 500 iterations. After 2500 iterations, the weights are well ordered with respect to the input data shown in Fig. 5.7.

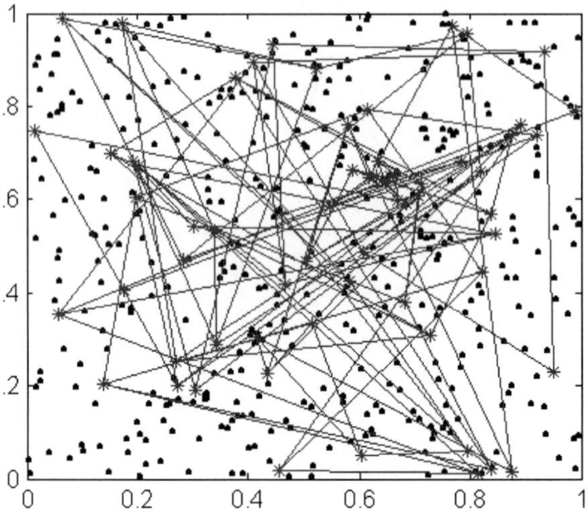

Figure 5.5. Initial weight distributions with neighborhood connections between neighboring neurons (Map size 8×8). Black dots are input data. Red stars are weights of neurons. Red lines are connections of neighboring neurons

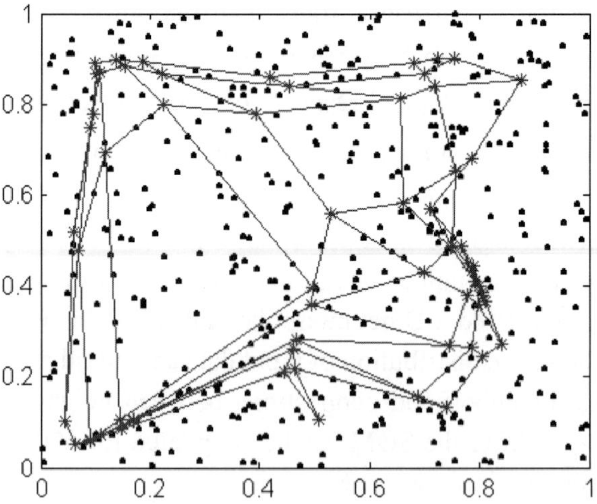

Figure 5.6. Weight distributions after 500 iterations. Black dots are input data; Red stars are weights of neurons; Red lines are connections of neighboring neurons

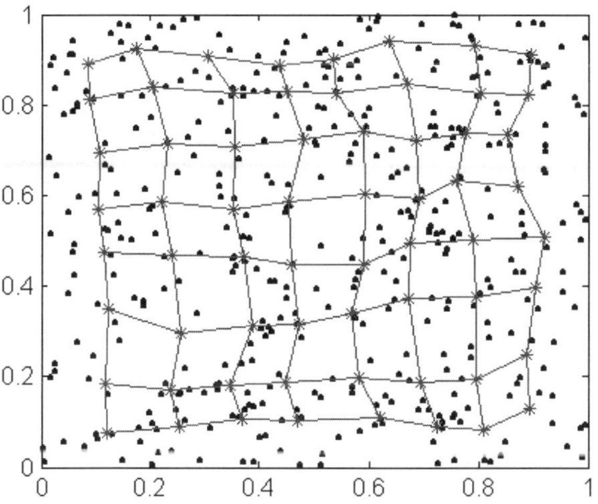

Figure 5.7. Weight distributions after training (2500 iterations). Black dots are input data; Red stars are weights of neurons; Red lines are connections of neighboring neurons

5.3. Growing SOMs

5.3.1. Cell Splitting Grid

This section describes a Cell Splitting Grid (CSG) learning algorithm for growing an SOM. The CSG algorithm improves the learning mechanism by growing an SOM in both the input and output space. It resembles the cell-splitting mechanism from a biological perspective. Using the CSG algorithm, it can overcome some of the shortcomings of conventional SOM. The CSG algorithm has some connections to quadtree structure, and other supervised self-organizing networks. It improves the performance for non-uniformly distributed input data. It enables a 2-D representation on the output map confined in a square region and neurons are distributed on a 2-D map according to the density distribution of input data. The neurons representing the dense region of input data are densely distributed on a 2-D output map, whereas those lying in the sparse region are located on the sparse region of the 2-D output space. As a result, the non-uniform distribution of neurons on the output map is not only able to preserve the data distribution in the input space, it also delivers a better performance of vector quantization than those delivered by conventional SOM and other SOM related algorithms.

The CSG network architecture is similar to conventional SOM architecture. Its output map is constrained in a square of unit length. All neurons are generated within the square. Each neuron corresponds to a square region with different size and neighboring neurons are connected to form the neighboring relationship. The notation of the CSG algorithm is introduced as follows. In order to decide when new neurons are generated, τ is introduced to denote the activation level of neurons. *XY* is the 2-D coordinates of a neuron in a 2-D square region. *Len* denotes the length of the square region. *L, R, T* and *B* denote the left, right, top and bottom bound coordinates of the square region. When new neurons are generated, they are all endowed with an initial value τ_i as the activation level τ. When a neuron is activated, its activation level τ decreases by a constant value. This process continues until τ of a neuron becomes zero. The neuron is then split to generating its four offspring neurons. The executing steps of the CSG algorithms are as follows.

1. Start from only one neuron and initialize its weight w with a random value within the input space. Set *L, R, T, B, Len* to 0, 1, 1, 0 and 1 respectively and set *XY* to [0.5 0.5]. Physically, this implies the first neuron be at the center of a square with a unit length. The square denotes the region corresponding to that neuron. Set τ of the first neuron to τ_i large enough to learn the information before splitting. *Cycle*, which is denoted as the iterations performed on the network before splitting, is set to zero.
2. Select an input x randomly from the input space according to the input distribution $p(x)$.
3. Find the best-matching neuron c by Eq. (5.1).
4. Adapt weights of the winner neuron c and neurons in N_c.

$$\Delta w_c = \alpha_1 (x - w_c)\text{, for the winner neuron c} \tag{5.2}$$

$$\Delta w_b = \alpha_2 (x - w_b)\text{, for all } b \in N_c, \alpha_2 \ll \alpha_1 \tag{5.3}$$

5. Decrease the activation level of the winner neuron c and increase the iteration counter:

$$\Delta\tau_c = -1 \tag{5.4}$$

$$\Delta Cycle = 1 \tag{5.5}$$

6. When the activation level τ of the current winner neuron c decreases to zero, perform the cell-splitting mechanism, i.e., delete the neuron c and then generate four new neurons within the square region of the neuron c, and set *Cycle* zero.
7. Initialize the new weights and the activation levels of the new generated neurons. The new weights are endowed according to the weight distribution on the output map before splitting. After the initial activation levels (τ_i) are given to the new neurons, increase the activation levels of all neurons by $\Delta\tau$ (>0) to slow down the splitting rate.
8. If *Cycle* is less than the pre-defined saturated time value *TMax*, then go to step 2. Otherwise, the neural network approaches to a saturate state indicating that an appropriate network topology is obtained.

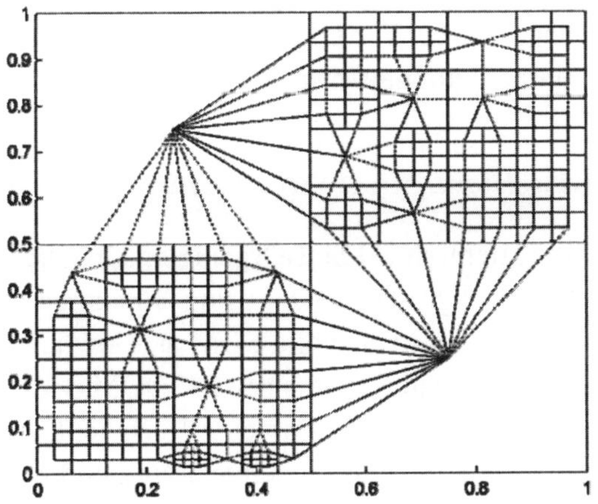

(a)

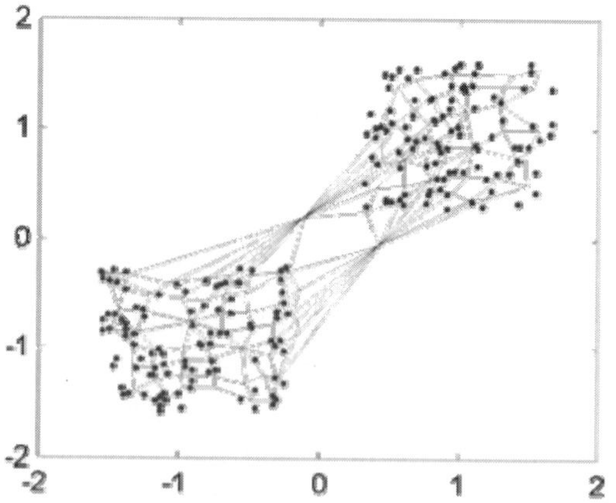

(b)

Figure 5.8. Learning results by the CSG algorithm (neighboring neurons are connected) (a) Output map (b) input data and neurons in the input space

Figure 5.8 shows an example of the CSG algorithm on a 2-D synthetic input data set. Neurons at the corners of the output map are dense. The two largest regions are located at the other two corners of the output map, representing the two neurons lying in the gap between the two dense regions in the input space. The CSG algorithm is useful in delving the data for determining the input data distribution.

5.3.2. Growing Hierarchical Self-Organizing Quadtree Map

The Growing Hierarchical Self-Organizing Quadtree Map (GHSOQM) is another hierarchical growing SOM using a quadtree structure. A neuron at a high level can generate its child SOM at a low level according to the number of inputs associated to it. The GHSOQM does not grow neurons horizontally for maintaining a simple and efficient growing process. A neuron in the GHSOQM may generate four child neurons upon some conditions. Fig. 5.9(a) shows the typical structure of an GHSOQM, which is very similar to the quadtree structure. The number of neurons at each level of the GHSOQM is adaptively determined. If we look down from the top of the GHSOQM and only use the leaf neurons that have no child SOMs, the final one-layer map is like the one-layer quadtree-like SOMs as shown in Fig. 5.9(b) corresponding to Fig. 5.9(a) Usually the root level with only one neuron is useless and we begin with the first level with 4 neurons.

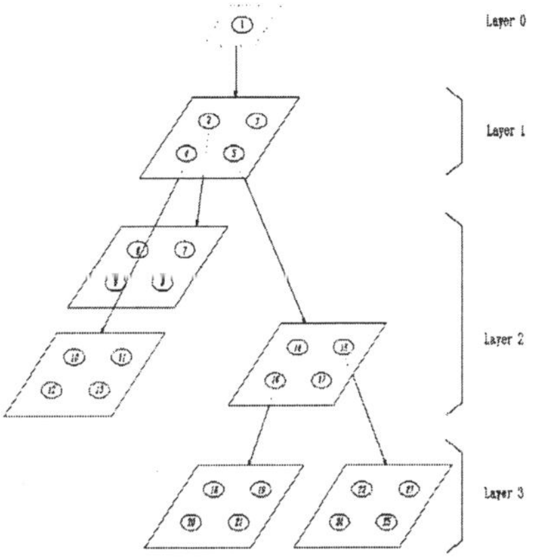

(a)

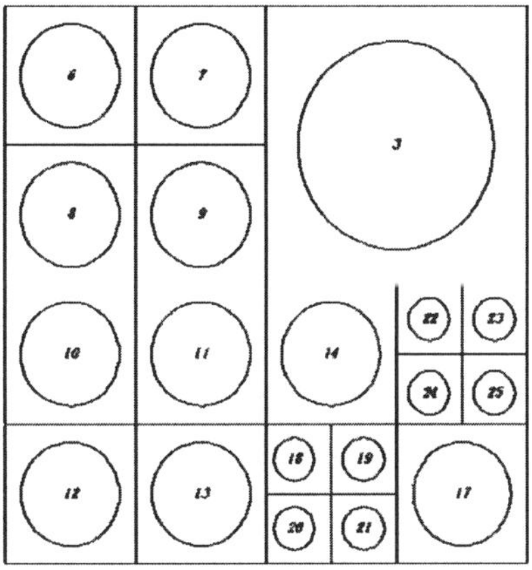

(b)

Figure 5.9. (a) Architecture of an GHSOQM that grows neurons hierarchically (b) The one-layer map corresponding to (a) when we look down from the top of hierarchy of the GHSOQM and only use the leaf neurons without child SOMs

The training data set is denoted as $X=\{x_1,\cdots,x_n\}$. The input data associated with neuron *i* at the *n*th level are denoted by $X_n(i)$. The feature vectors of the child neurons at the (*n+1*)th level from the mother neuron *i* at the *n*th level is denoted by $W_{n+1}^i=\{W_{n+1}^i(1),W_{n+1}^i(2),W_{n+1}^i(3),W_{n+1}^i(4)\}$. The GHSOQM algorithm is summarized as follows.

1. INITIALIZATION:
   Set level n=1 and the feature vectors at the first level $W_1^0=\{W_1^0(1),W_1^0(2),W_1^0(3),W_1^0(4)\}$, where $W_1^0(i)$ is the feature vector of the ith neuron at level 1. An SOM with the four neurons is trained with all data X by invoking the function TRAIN_SOM (X, n, W_1^0).

2. RECURSIVE LOOP:
   FUNCTION GENERATE_SOM (X, n, W)
   FOR i=1 to 4
      Assign each input datum in X to its nearest neurons. If the number of inputs assigned with the ith neuron at the nth level is more than a predefined number τ, then the neuron spawns four child neurons representing a child SOM with size 2×2. Then train the child SOM by TRAIN_SOM($X_n(i)$, n+1, W_{n+1}^i) and generate child SOMs by recursively invoking GENERATE_SOM($X_n(i)$, n+1, W_{n+1}^i).
   END

   FUNCTION TRAIN_SOM (INPUT, n, W)
   Train SOM at the nth level with the input data INPUT and the four neurons with feature vectors W.

The function *TRAIN_SOM* is an implementation of classical SOM algorithm. The function *GENERATE_SOM* recursively generates child SOMs if possible and train them with data associated with their mother neurons. In a word, the GHSOQM trains the network at each level by the data associated with their mother neurons. The GHSOQM completes the training from the upper levels and then proceeds to train the next lower level. For newly added data, the learned network must have the ability to learn new data without reusing old data.

5.4. Probabilistic SOMs

5.4.1. Cellular Probabilistic SOM

The SOM can be considered as a combination of vector quantization (VQ) and topology preservation. If no lateral interactions occur, an SOM becomes a standard VQ algorithm. Through adding topological information into the cost function in VQ problems, an SOM can be viewed as an efficient approximation to the gradient descent algorithm for topological VQ (TVQ). The TVQ can be further extended to Soft Topographic Vector Quantization (STVQ), which provides a soft probability assignment for each neuron. Using a batch mode Expectation-Maximization (*EM*) algorithm, the STVQ offers a family of topographic mapping algorithms in which the batch SOM and TVQ are included. The STVQ algorithm gives an energy function (soft quantization error) as follows.

$$F = \sum_{t=1}^{M} \sum_{i=1}^{N} P_i(x(t)) \sum_{j=1}^{N} h_{ij} D(x(t), w_j) \tag{5.6}$$

where M and N are the numbers of input data and map size respectively, $P_i(x(t))$ is the soft assignment probability of neuron i upon the input $x(t)$, h_{ij} is a fixed neighborhood function satisfying $\sum_{j=1}^{N} h_{ij} = 1$, and $D(x(t), w_j)$ is the quantization error between the input $x(t)$ and the weight w_j of neuron j, defined by $D(x(t), w_j) = \frac{1}{2}\|x(t) - w_j\|^2$. The entropy of the assignments is

$$S = -\sum_{t=1}^{M} \sum_{i=1}^{N} P_i(x(t)) \ln(P_i(x(t)) \tag{5.7}$$

In order to maximize the entropy of Eq. (5.8) with the constraint given in Eq. (5.7), the energy function becomes

$$E = \beta F - S \tag{5.8}$$

Using the *EM* algorithm, the weights can be obtained by the following iterative steps:

$$\text{E step: } P_i(x(t)) = \frac{\exp(-\beta \sum_{j=1}^{N} h_{ij} D(x(t), w_j))}{\sum_{k=1}^{N} \exp(-\beta \sum_{j=1}^{N} h_{kj} D(x(t), w_j))}, i = 1,\ 2,\ \ldots,\ N \tag{5.9}$$

$$\text{M step: } w_i = \frac{\sum_{t=1}^{M} x(t) \sum_{j=1}^{N} h_{ij} P_j(x(t))}{\sum_{t=1}^{M} \sum_{j=1}^{N} h_{ij} P_j(x(t))}, i = 1, 2, \ldots, N \tag{5.10}$$

In Eq. (5.9), β is a parameter of inverse temperature. The optimized weights can be obtained using deterministic annealing from low to high values in order to avoid being stuck at the local minima of the energy function of Eq. (5.9). The above steps of the STVQ algorithm are batch type and can be modified into the batch type SOM if we set h_{ij} to a delta function, δ_{ij}, and $\beta \to \infty$ in Eq. (5.9). As it is needed to collect all the input data before the learning process, the batch EM algorithm is not suitable for large data sets or dynamic data sets. Thus, an online *EM* algorithm is necessary to be incorporated in the STVQ model.

We describe an online mode of STVQ, called the Cellular Probabilistic SOM (CPSOM). The term "cellular" is derived from the concept that a neuron in the CPSOM is locally connected to neighboring neurons like a biological cell. However, the incremental *EM* algorithms are not suitable for the CPSOM because they need additional storage variables for all the training data. The online EM model is able to improve incremental EM algorithms when the parameters in the *M* step can be expressed by sufficient statistics. This leads to the derivation of the CPSOM. The online form of CPSOM is equivalent to the STVQ under certain conditions. The online property makes it more suitable for large data sets than the STVQ. It shows a faster convergence rate than the STVQ with the same effect when map size is relatively small. When the map size becomes relatively large, the stochastic approximation to the STVQ in turn makes it less likely for the CPSOM be trapped in local minima, thereby the final maps are more ordered with lower soft quantization errors than the STVQ. Most of all, the online CPSOM can be used for dynamic data sets. It has relations to the SOM, S-Map, and the online gradient-descent form of STVQ.

We can express the weight-updating in the *M* step as an online version. First, we introduce a state variable $B_i(t)$ for neuron i at iteration t:

$B_i(t) = \sum_{k=1}^{t}\sum_{j=1}^{N} h_{ij}P_j(x(k))$, then the weight of neuron i can be sequentially adapted by

$$w_i(t) = \frac{w_i(t-1)B_i(t-1) + x(t)\sum_{j=1}^{N} h_{ij}P_j(x(t))}{B_i(t-1) + \sum_{j=1}^{N} h_{ij}P_j(x(t))} \tag{5.11}$$

$$w_i(t) = w_i(\text{t-1}) + \Delta w_i$$

$$\Delta w_i = \frac{1}{B_i(t)}\sum_{j=1}^{N} h_{ij}P_j(x(t))(x(t)-w_i(t-1)),\quad i=1,\ 2,\ \ldots,\ N \tag{5.12}$$

The learning rate η_i in Eq. (5.12) is $\frac{1}{B_i(t)}$. The updating rule in Eq. (5.12) is also intended to optimize the objective function of Eq. (5.9).

The CPSOM algorithm can be implemented as the following steps:

1. Initialize weights of all neurons to be random vectors, compute the neighborhood function h_{ij}, and set the start values $\beta = \beta^{start}$ and $\gamma = \gamma^{start}$.
2. Select an input x randomly from a static data set, or sequentially from a dynamic data set.
3. Adapt the weights and state variables for all neurons:

$$E\ \text{step: } P_i(x(t)) = \frac{\exp(-\beta\sum_{j=1}^{N} h_{ij}D(x(t), w_j))}{\sum_{n=1}^{N}\exp(-\beta\sum_{j=1}^{N} h_{nj}D(x(t), w_j))},\ i=1,\ 2,\ \ldots,\ N$$

$$M \text{ step: } w_i(t) = w_i(t-1) + \frac{1}{B_i(t)} \sum_{j=1}^{N} h_{ij} P_j(x(t))(x(t) - w_i(t-1)),\ i = 1, 2, \ldots, N$$

where $B_i(t) = B_i(t-1) + \sum_{j=1}^{N} h_{ij} P_j(x(t)), i = 1, 2, \ldots, N$.

4. Increase β by $\beta \leftarrow \beta + \Delta_1$ if $\beta < \beta^{final}$, where $\Delta_1 (> 0)$ is an incremental factor, if the current iteration t is an integer multiple of a parameter λ_1.
5. If the current iteration t is an integer multiple of a parameter λ_2, decrease $B_i(t)$ by $B_i(t) = B_i(t)/\gamma,\ i = 1, 2, \ldots, N$, where $\gamma \leftarrow \gamma - \Delta_2 (\gamma \geq 1, \Delta_2 > 0)$ is a factor shrinking to a final value $\gamma^{final} = 1$. This procedure enables the algorithm to retrain the network at a higher learning rate after certain iterations to avoid being trapped in local minima. In fact, γ can be considered as a forgetting factor to reduce the weight of the old input data contained in the $B_i(t)s$. When an CPSOM is used for dynamic data sets, we set $\gamma = \gamma^{start}$ if the current iteration t is an integer multiple of a parameter λ_3 ($\lambda_3 >> \lambda_2$ and λ_3 is an integer multiple of λ_2). This procedure ensures the learning algorithm to track the dynamic environment.
6. Terminate the adaptation for static data sets when pre-specified epochs T^{final} are reached. Otherwise, go to *step* 2.

It should be noted that the parameter λ_3 is not required if the input data are static. The learning rate is oscillatory at the early stage of the training and finally decreases to zero. For dynamic data sets, γ needs to be reset to γ^{start} to retrain the network after sufficient iterations. This requires that the models under the dynamic data do not change too rapidly.

5.4.2. *Probabilistic Regularized SOM*

The ViSOM is a relatively recent algorithm for preserving topology as well as inter-neuron distances. The final map can be seen as a smooth net embedded in input space. The distances between any pairs of neurons in input space resemble those in output space. The ViSOM uses the same network architecture of the SOM. The only difference between the two networks is that the neighboring neurons of winner neuron are updated differently. The weight-updating rule (5.2) is used for SOM, while the weight-updating rule for the neighboring neurons of winner neuron in the ViSOM is

$$w_i(t+1) = w_i(t) + \varepsilon(t) h_{ic}(t) \left([x(t) - w_c(t)] + [w_c(t) - w_i(t)] \left(\frac{d_{ci} - \lambda\Delta_{ci}}{\lambda\Delta_{ci}} \right) \right), \quad \forall i \in N_c \tag{5.13}$$

where d_{ci} and Δ_{ci} are the distances between the neuron c and i in input space and output space, respectively, and λ is a resolution parameter. The basic idea behind the ViSOM is that the force $F_{ix} = x(t) - w_i(t)$ can be decomposed into two parts: $F_{ix} = [x(t) - w_c(t)] + [w_c(t) - w_i(t)] = F_{cx} + F_{ci}$. F_{cx} is a force from the winner neuron c to the input x. F_{ci} is a lateral force from the neuron i to the winner neuron c. The ViSOM constrains the lateral force F_{ci} by multiplying a coefficient $\frac{d_{ci} - \lambda\Delta_{ci}}{\lambda\Delta_{ci}}$. The objective is to maintain the preservations of distances between neurons. The discrete surface constructed by neurons is then regularized to be smooth for good visualization. In order to keep the rigidity of the final map, the final neighborhood size σ_f should not include only the winner neurons. The larger σ_f, the flatter the map in input space will be. The resolution parameter λ controls the resolution of the map. Small values of λ generate maps with high resolution, while large values of λ generate maps with low resolution.

Here we describe the Probabilistic Regularized SOM (PRSOM) algorithm. Unlike the hard assignment in the SOM and ViSOM, the assignment in the PRSOM is soft so that an input datum belongs to a neuron with certain probability. The sequential weight-updating rule of the PRSOM is extended from the ViSOM to an optimization of a cost function. Under certain circumstance, the ViSOM can be considered as a special case and an accelerated one of PRSOM. The PRSOM can also be considered as a discrete approximation of principal surfaces like the SOM and ViSOM. Like regularization terms used in supervised learning, quantization and feature extraction to simplify or smooth function and to avoid overfitting, the surfaces of PRSOM are smooth that enables a good visualization effect.

Let $p_j(x(t))$ denote the noised probabilistic assignment of neuron j:

$$p_j(x(t)) = \sum_{i=1}^{N} h_{ij} P_i(x(t)) \tag{5.14}$$

where $P_j(x(t))$ is the probabilistic assignment of neuron j for input $x(t)$ and h_{ij} is a neighborhood constant satisfying $\sum_{j=1}^{N} h_{ij} = 1$. Here the term "noised" mean $p_j(x(t))$ is affected by leaked probabilistic assignments from other neighboring neurons. Therefore $p_j(x(t))$ is the probabilistic assignment of neuron j that considers the effects of other neurons. Note that $p_j(x(t))$ can be considered as a weight since

$$\sum_{j=1}^{N} p_j(x(t)) = \sum_{j=1}^{N}\sum_{i=1}^{N} h_{ij} P_i(x(t)) = \sum_{i=1}^{N} P_i(x(t)) \sum_{j=1}^{N} h_{ij} = 1 \tag{5.15}$$

The cost function of the PRSOM is then soft vector quantization error:

$$F_{vq} = \frac{1}{2}\sum_{t=1}^{M}\left\| \sum_{j=1}^{N} p_j(x(t))\left[x(t) - w_j\right]\right\|^2 = \frac{1}{2}\sum_{t=1}^{M}\left\| x(t) - \sum_{j=1}^{N} p_j(x(t)) w_j \right\|^2 \tag{5.16}$$

which computes the sum of square errors between the input data and the average weights for all input data. To control the complexity of the above model, or ensure the solution simple or smooth, we added the following regularization term:

$$F_{reg} = \frac{1}{8}\sum_{t=1}^{M}\sum_{j=1}^{N}\sum_{m=1}^{N} p_j(x(t))p_m(x(t))(d_{jm}{}^2 - \lambda\Delta_{jm}^2)^2/(\lambda\Delta_{jm}^2 + I_{jm}) \quad (5.17)$$

where $d_{jm} = \|w_j - w_m\|$ is the distance in input space, Δ_{jm} is the corresponding distance between neuron j and m in 2-D output space, λ is a resolution parameter like the ViSOM, and the identity matrix I is introduced to avoid the case that the denominator of the fractional term would be zero when $j=m$. F_{reg} in (5.17) tries to preserve pairwise distance of neurons in input and output space. It emphasizes large products of errors and fractional errors like Sammon's mapping. It also can be considered as the restriction of the PRSOM for the smoothness of discrete approximation of the principal surfaces. Then a regularized cost function of the PRSOM is

$$E = F_{vq} + \gamma F_{reg} \frac{1}{2}\sum_{t=1}^{M}\left\|\sum_{j=1}^{N} p_j(x(t))\left[x(t) - w_j\right]\right\|^2 + \frac{\gamma}{8}\sum_{t=1}^{M}\sum_{j=1}^{N}\sum_{m=1}^{N} p_j(x(t))p_m(x(t))(d_{jm}{}^2 - \lambda\Delta_{jm}^2)^2/(\lambda\Delta_{jm}^2 + I_{jm}) \quad (5.18)$$

where γ is a regularization parameter.

The weight-updating and probability assignment of the PRSOM can be explained from Eq. (5.18) and can be re-expressed as $E = \sum_{t-1}^{M} E(t)$ where:

$$E(t) = \frac{1}{2}\left\|\sum_{j=1}^{N} p_j(x(t))\left[x(t) - w_j\right]\right\|^2 + \frac{\gamma}{8}\sum_{j=1}^{N}\sum_{m=1}^{N} p_j(x(t))p_m(x(t))(d_{jm}{}^2 - \lambda\Delta_{jm}^2)^2/(\lambda\Delta_{jm}^2 + I_{jm})$$

Since the left and right terms in $E(t)$ are always positive, the minimization of E is equal to the minimization of each $E(t)$.

Taking the gradient of $E(t)$ with respect to w_j, i.e.,

$$\frac{\partial E(t)}{\partial w_j} = -p_j(x(t))\sum_{i=1}^{N} p_i(x(t))(x - w_i)$$

$$-\gamma p_j(x(t))\sum_{i=1}^{N} p_i(x(t))(w_i - w_j)\left(\frac{d_{ij}^2 - \lambda\Delta_{ij}^2}{\lambda\Delta_{ij}^2 + I_{ij}}\right)$$

the following weight-updating rule is obtained:

$$w_j(t+1) = w_j(t) - \varepsilon(t)\frac{\partial E(t)}{\partial w_j}$$

$$= w_j(t) + \varepsilon(t)p_j(x(t))\left[\sum_{i=1}^{N} p_i(x(t))\left([x(t) - w_i(t)] + \gamma[w_i(t) - w_j(t)]\left(\frac{d_{ij}^2 - \lambda\Delta_{ij}^2}{\lambda\Delta_{ij}^2 + I_{ij}}\right)\right)\right] \tag{5.19}$$

In (5.19), $\varepsilon'(t) = \varepsilon(t)p_j(x(t))$ is the learning rate of the weight-updating rule of the PRSOM. To avoid small values of the learning rate $\varepsilon'(t)$, the noised probabilistic assignment (or fuzzy neighborhood function) $p_j(x(t)) = \sum_{i=1}^{N} h_{ij}(t)P_i(x(t))$ can be set $p_j'(x(t)) = p_j(x(t))/\max_k\left(p_k(x(t))\right)$. Then the resultant updating rule is

$$w_j(t+1) = w_j(t) + \varepsilon(t)p_j'(x(t))$$

$$\left[\sum_{i=1}^{N} p_i(x(t))\left([x(t) - w_i(t)] + \gamma[w_i(t) - w_j(t)]\left(\frac{d_{ij}^2 - \lambda\Delta_{ij}^2}{\lambda\Delta_{ij}^2 + I_{ij}}\right)\right)\right] \tag{5.20}$$

The probabilistic assignment $P_i(x(t))$ is (5.10) in the STVQ. But the additional parameter β, the inverse temperature, must be carefully selected and tuned from low to high values. If we used the same technique in the PRSOM, we add the entropy into the cost function (5.18):

$$E = \beta(F_{vq} + \gamma F_{stress}) + \sum_{t=1}^{M}\sum_{i=1}^{N} P_i(x(t))\ln(P_i(x(t))) \tag{5.21}$$

where β is a fixed regularization parameter. Taking the gradient of (5.21) with respect to $P_j(x(t))$, we obtained the following expression of $P_j(x(t))$:

$$P_j(x(t)) = \frac{\exp\left\{-\beta\left\{\begin{array}{l}\left(\sum_{k=1}^{N}\sum_{i=1}^{N} h_{ik}P_i(x(t))(x(t)-w_k)\right)^T [x(t)-\sum_{i=1}^{N} w_i h_{ji}] \\ +\gamma\sum_{k=1}^{N}(\sum_{i=1}^{N} h_{ik}P_i(x(t)))\sum_{m=1}^{N} h_{jm}(d_{mk}{}^2-\lambda\Delta_{mk}^2)^2/(\lambda\Delta_{mk}^2+I)\end{array}\right\}\right\}}{\sum_{n=1}^{N}\exp\left\{-\beta\left\{\begin{array}{l}\left(\sum_{k=1}^{N}\sum_{i=1}^{N} h_{ik}P_i(x(t))(x(t)-w_k)\right)^T [x(t)-\sum_{i=1}^{N} w_i h_{ni}] \\ +\gamma\sum_{k=1}^{N}(\sum_{i=1}^{N} h_{ik}P_i(x(t)))\sum_{m=1}^{N} h_{nm}(d_{mk}{}^2-\lambda\Delta_{mk}^2)^2/(\lambda\Delta_{mk}^2+I)\end{array}\right\}\right\}} \tag{5.22}$$

which is a fixed-point iteration. However, Eq. (5.22) may not converge in practical situations. A more convenient and heuristic way to compute $P_i(x(t))$ can be taken as

$$P_j(x(t)) = \frac{1}{C}\left(1/\left\|\sum_{k=1}^{N} h_{jk}\left(x(t)-w_k\right)\right\|^2\right) \tag{5.23}$$

where C is a normalization constant. No iteration is needed in (5.23). Since neighborhood function h_{jj} is much larger than any other $h_{jk}(k \neq j)$, $P_j(x(t))$ achieves the highest probability assignment if w_j is the feature vector of the nearest neuron from the input $x(t)$. Eq. (5.23) is then reasonable in that the closer a neuron to input, the higher assignment probability.

The architecture of the PRSOM is the same as the SOM or ViSOM. Using the same notation of SOM, the sequential PRSOM algorithm is described as follows.

1. Randomly select an input $x(t)$ from a data set.
2. Compute the assignment probability of $x(t)$ for all neurons according to Eq. (5.23).

3. Perform the weight-updating rule for all neurons according to Eq. (5.20).
4. Terminate the algorithm until certain criterion is satisfied. Otherwise, go to step 1.

The above sequential algorithm is affected by the ordering of training samples. To avoid this problem, it is better to use the following batch algorithm of PRSOM.

1. Compute the assignment probability of $x(t)$ for all input data and neurons according to Eq. (5.23).
2. Perform the batch weight-updating rule for all neurons:
$$w_j(k+1) = w_j(k) + \frac{\varepsilon(t)}{N}\sum_{t=1}^{N} p_j'(x(t))$$
$$\times\left(\sum_{i=1}^{N} p_i(x(t)\left([x(t)-w_i(k)]+\gamma[w_i(k)-w_j(k)]\left(\frac{d_{ij}^2-\lambda\Delta_{ij}^2}{\lambda\Delta_{ij}^2+I_{ij}}\right)\right)\right)$$
where k is current epoch, k+1 is the next epoch.
3. Terminate the algorithm until certain criterion is satisfied.

The computational complexity of the PRSOM and ViSOM is $O(MN^2)$, where M and N are the number of input data and neurons, respectively. If N^2 is significantly less than M, the computational complexity of the PRSOM is less than that of MDS, i.e., $O(M^2)$. $\varepsilon(t)$ in the PRSOM should be decreased from high values to nearly zero. The selection of regularization coefficient γ can be set from 0.5 to 10 practically according to emphasis of the second term in Eq. (5.18). h_{ij} in the PRSOM can be selected in Eq. (5.24) with the constraint $\sum h_{ij}=1$:

$$h_{ij} = \exp\left(-\frac{\|Pos_i - Pos_j\|^2}{2\sigma^2}\right) \Bigg/ \sum_{k=1}^{N}\exp\left(-\frac{\|Pos_i - Pos_k\|^2}{2\sigma^2}\right) \tag{5.24}$$

where the neighborhood radius σ is a constant. The value of σ is important for the training. The neighborhood function curves are steep when the value of σ is small. As a result, only few neurons around neuron

j can be included in the computation of the weight of the neuron j. This may have an effect of generating folded or disordered maps. On the other hand, the area of the neighborhood function is enlarged to neurons that are far from the neuron j when σ is set to a large value. Large σ flattens the neighborhood function curves and results in contracted maps. This would degrade the performance of competitive learning. Generally, it is suggested to set σ to 0.5 which results in maps with good mapping effects.

The neighborhood function in the PRSOM is $p_j(x(t)) = \sum_{i=1}^{N} P_i(x(t))h_{ij}(t)$. σ in Eq. (5.24) can be set to a small value, e.g., 0.5, such that $p_j(x(t))$ affects not only the winner neuron due to the leaked information from other neighboring neurons. The most important property of the PRSOM is the cost function in Eq. (5.18), which gives the meaning of the weight-updating rule. From the definition of the cost function, the probabilistic quantization error F_{vq} in Eq. (5.16) is different from that of the STVQ in Eq. (5.7). Only optimizing the first term F_{vq} will not generate similar result with the SOM. This should be also true for the ViSOM if the regularized term in the updating rule is left out. The implication of F_{vq} is to not only minimize the probabilistic quantization error, but also repulse neurons from one another.

The resolution parameter λ must be chosen carefully. If λ is too large, some useful data structure may not be well displayed on the output map. Some neurons far outside input data may be wasteful for visualization. If λ is too small, the resultant map is embedded in input data and cannot well display input data. The following show a practical equation for the selection of λ:

$$\lambda = 1 \sim 1.5 \times \frac{Span_{\max}}{\min\{a,b\}}, \text{ or } \lambda = 1 \sim 1.5 \times \frac{4 \times \sqrt{Var_{\max}}}{\min\{a,b\}} \tag{5.25}$$

where a and b are the number of rows and columns of the SOM map, respectively. However, the selection of λ may be out of the rage according to Eq. (5.25) because of high input dimension or nonlinearity. The soft assignment in the PRSOM can be exploited similar to that in the

STVQ. The accumulated probability in each neuron forms an accumulated probability matrix (*AP* matrix) like U-matrix. The element AP_{ij} of neuron k located at the ith row and j^{th} column of the map is defined by:

$$AP_{ij} = \sum_{t=1}^{M} P_k(x(t)) \tag{5.26}$$

Assigning different colors to different accumulated probabilities, we can obtain a colored map with some colors corresponding to clusters and some colors corresponding to empty regions. This is a powerful visualization technique in addition to the method by simply assigning input data to their nearest neurons. The SOM and ViSOM are both discrete approximations of principal surfaces. But SOM cannot display the data boundary well at the boundaries of output map because it is a density-based quantizer. The ViSOM instead can represent the data boundary well because ViSOM is a uniform quantizer and some neurons are outside input data when the parameters of an ViSOM are properly chosen. The PRSOM is also a discrete approximation of principal surface like the ViSOM. As the inter-neuron distances in input space are regularized to resemble those in output grid, the regularized term (second term) in Eq. (5.19) can be very small after the completion of training. We further consider only the nearest neuron c using hard assignment. The updating rule in Eq. (5.19) now becomes

$$w_j(t+1) = w_j(t) + \varepsilon(t) h_{cj} \left(x(t) - w_c(t) \right) \tag{5.27}$$

Then the adaptation rule in the final stage leads to the smoothing process:

$$\text{PRSOM: } w_c = \sum_{t=1}^{L} x(t) h_{cj} \Big/ \sum_{t=1}^{L} h_{cj} \tag{5.28}$$

where h_{cj} is fixed for all time in the PRSOM.

5.5. Clustering of SOM

The clustering at the second level for an SOM-based two-level clustering is the agglomerative hierarchical clustering. The merging criterion is motivated by a clustering validity index based on the inter-cluster and intra-cluster density, and inter-cluster distances (Halkidi & Vazirgiannis, 2002). The original index is used for the whole input data and therefore is a global index. The optimal number of clusters can be found by the clustering validity index. In this SOM based clustering algorithm (Wu & Chow, 2004), the clustering validity index is slightly modified and used locally to determine which neighboring pair of clusters to be merged into one cluster in the agglomerative hierarchical clustering. Since more information is added into the merging criterion in addition to the inter-cluster distances, the described algorithm clusters data in a better way than other clustering algorithms using an SOM. After certain preprocessing techniques for filtering, the clustering algorithm is able to handle the input data with noises and outliers.

The notations in the clustering validity index are defined as follows. The data set is partitioned into c clusters. A set of representation points $V_i = \{v_{i1}, v_{i2}, \ldots, v_{ir_i}\}$ represents the ith cluster, where r_i is the number of representation points of the ith cluster. *stdev*(i) is a standard deviation vector of the ith cluster. The pth component of *stdev*(i) is defined by $stdev^p(i) = \sqrt{\sum_{k=1}^{n_i} (x_k^p - m_i^p)^2 / (n_i - 1)}$, where n_i is the number of data points in the ith cluster, x_k is the data belonging to the ith cluster, and m_i is the sample mean of the ith cluster. The average standard deviation is given by: $stdev = \sqrt{\sum_{i=1}^{c} \|stdev(i)\|^2 / c}$.

The overall clustering validity index, called "Composing Density Between and With Clusters" (*CDbw*), is defined by

$$CDbw(c) = Intra\_den(c) \times Sep(c) \tag{5.29}$$

Intra_den(c) in Eq. (5.29) is the intra-cluster density and defined by

$$Intra\_den(c) = \frac{1}{c}\sum_{i=1}^{c}\sum_{j=1}^{r_i} density(v_{ij}), \qquad c > 1 \qquad (5.30)$$

where $density(v_{ij})$ is defined by $density(v_{ij}) = \sum_{l=1}^{n_i} f(x_l, v_{ij})$, where x_l belongs to the ith cluster, v_{ij} is the jth representation point of the ith cluster, and $f(x_l, v_{ij})$ is defined by

$$f(x_l, v_{ij}) = \begin{cases} 1, & \|x_l - v_{ij}\| \le stdev \\ 0, & \text{otherwise} \end{cases} \qquad (5.31)$$

The *Sep(c)* in Eq. (5.19) is the inter-cluster density and defined by

$$Sep(c) = \sum_{i=1}^{c}\sum_{\substack{j=1 \\ j \ne i}}^{c} \frac{\|close\_rep(i) - close\_rep(j)\|}{1 + Inter\_den(c)}, \qquad c > 1 \qquad (5.32)$$

where *close_rep(i)* and *close_rep(j)* are the closest pair of representations of the *i*th and *j*th clusters, and *Inter_den*(*c*) is defined by

$$Inter\_den(c) = \sum_{i=1}^{c}\sum_{\substack{j=1 \\ j \ne i}}^{c} \frac{\|close\_rep(i) - close\_rep(j)\|}{\|stdev(i)\| + \|stdev(j)\|} density(u_{ij}), \qquad c > 1$$

where u_{ij} is the middle point between the pair points close_rep(i) and close_rep(j), $density(u_{ij}) = \sum_{k=1}^{n_i + n_j} f(x_k, u_{ij})$, where x_k is the input vector belonging to the ith and jth clusters, and $f(x_k, u_{ij})$ is defined by

$$f(x_k, u_{ij}) = \begin{cases} 1, & \|x_k - u_{ij}\| \le (\ \|stdev(i)\| + \|stdev(j)\|\ )/2 \\ 0, & \text{otherwise} \end{cases} \qquad (5.33)$$

The overall clustering algorithm is summarized as follows:

1. Train input data using an SOM.
2. Preprocessing before clustering of an SOM.

3. Cluster of an SOM using the agglomerative hierarchical clustering. Find the local *CDbw* for all pairs of directly neighboring clusters and merge the two clusters with the lowest *CDbw*. Compute the global *CDbw* for all input data before the merging process until only two clusters exist, or merging cannot happen.
4. Find the optimal partition of input data according to the global *CDbw* for all the input data as a function of the number of clusters.

Wine data set is used as an example for demonstration. Wine data have 178 13-D data with known 3 classes. The numbers of data samples in the three classes are 59, 71 and 48, respectively.

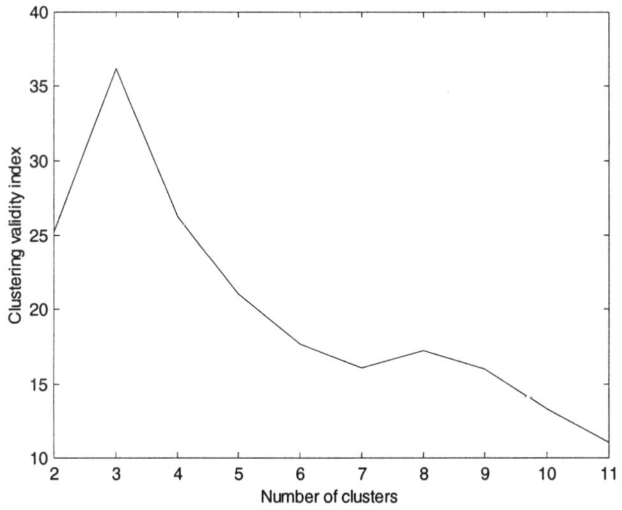

Figure 5.10. The *CDbw* as a function of the number of clusters for the wine data set by the clustering algorithm on SOM

We use this clustering algorithm with map size 4×4 to cluster the data. The *CDbw* as a function of the number of clusters, plotted in Figure 5.10, indicates that the number of clusters is three, which is exactly equal to the number of the classes. This algorithm achieved a high clustering accuracy of 98.3%.

5.6. Multi-Layer SOM for Tree-Structured data

The conventional SOM and the above mentioned SOM deal with vector type data of fixed length. They are, however, unable to deal with tree-structured data. A tree data can consists of many levels, and number of nodes in each level is not fixed. The Multi-Layer SOM (MLSOM) is another extended architecture developed for handling tree-structured data that cannot be encoded in a single fixed vector. The node attributes at different levels of trees are ordinal-type features. Using multiple SOM layers, the MLSOM processes the nodes of tree-structured data at different levels in a layer-by-layer fashion. Fig. 5.11 describes the schematic diagram of processing a 3-level tree data using the MLSOM.

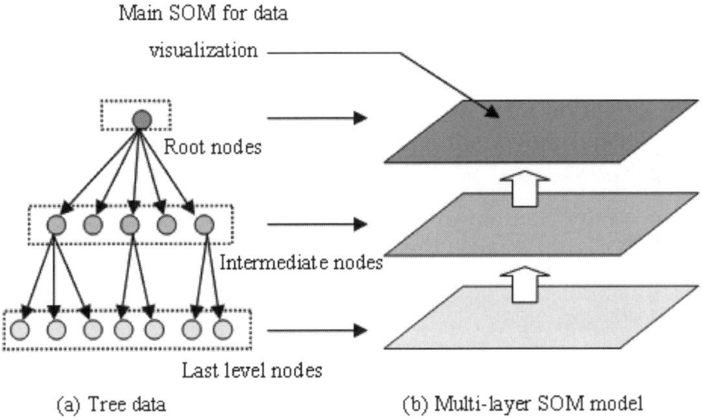

Figure 5.11. Data mapping in a multi-layer SOM model

The number of SOM layers is equal to the maximum levels of trees. If there are maximum L levels in all the tree structures, L groups of data and corresponding L SOM layers are generated. The ith (i=1,...,L) group of data is composed of all the ith level nodes in all the tree structures. The basic idea of the MLSOM is that the SOM training is performed in a way of level by level. That is, the Lth level group of data is firstly trained by the Lth SOM layer. Similar to the SOM-SD, the SOM outputs of child nodes are used for the input representation (child vector) of a parent node. After the Lth SOM output information is filled in the $(L-1)^{th}$ SOM input representation, the (L-1)th SOM layer is then trained. This procedure repeats until the 1st level group of data is trained by the 1st SOM layer. Finally, the visualization of tree-structured data can be performed on the 1st SOM layer.

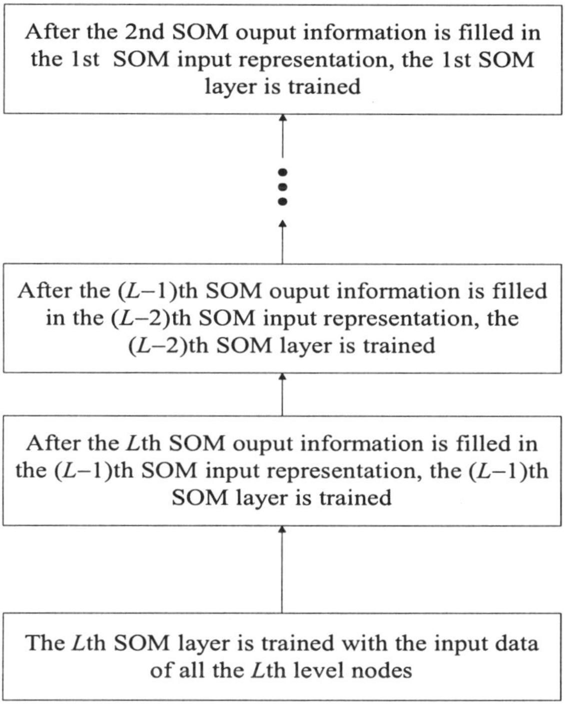

Figure 5.12. The basic flow chart of training steps by the MLSOM

The basic flow chart of training steps by the MLSOM is illustrated in Figure 5.12. The following describe the SOM input representation and the MLSOM training in details.

5.6.1. SOM Input Representation

Assume that the maximum number of children nodes of a node at the kth level in all the training data trees are c_k, the maximum levels in all the tree structures is L and the SOM output is a 2-D rectangular grid. It is noted that $c_L = 0$ as leaf nodes at the bottom layer have no children nodes. Each SOM input representation at an SOM layer consists of two parts: (1) the n-dimensional feature vector u of the current node; (2) 2-D position vectors $p_1,\ldots,p_{c_k}$ of its children nodes at the SOM layer below the current SOM. For non-leaf nodes, some p_i may be set to [0,0] since the number of children nodes of a node may be less than c_k. The lowest values of horizontal or vertical positions on the 2-D SOM output map are set to be larger than 0.

The ordering of children nodes of a node may not be predefined in real world applications. In the MLSOM, an ordering algorithm is used if the ordering of children is not predefined. Suppose that the nodes at the kth level need to be ordered before appending their position vectors to the SOM input representation at the $(k\text{-}1)^{th}$ SOM layer. After the completion of training the kth SOM layer, all the 2-D output positions of the nodes at k^{th} level are obtained.

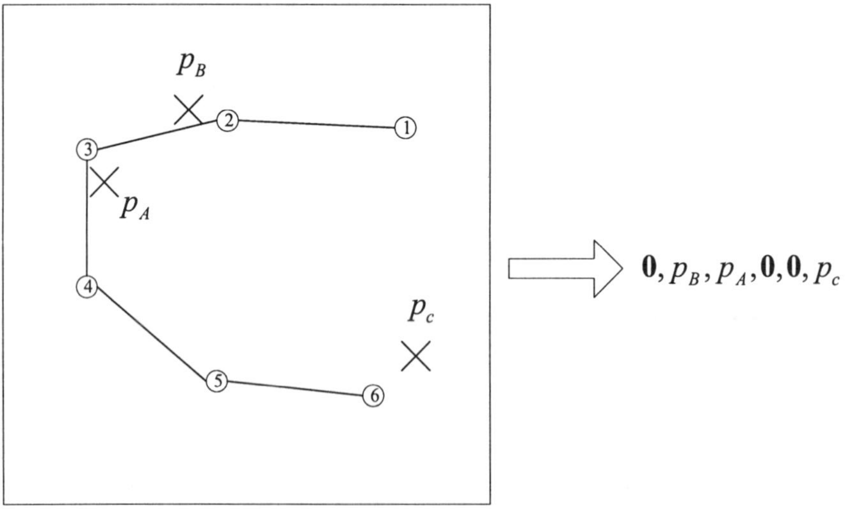

Figure 5.13. The illustration of the ordering of children nodes of a node

The basic idea of the ordering algorithm is to use all 2-D position vectors to train a 1-dimensional SOM. The number of neurons in the 1-D SOM is set to c_{k-1}, i.e., the maximum number of children nodes of a node at the $(k\text{-}1)^{th}$ level. After the completion of training of the 1-D SOM, each training datum is assigned to a neuron index of the 1-D SOM. The neuron index is then used in the SOM input representation of p_i ($i \in \{1,...,c\}$) of parent nodes at the $(k\text{-}1)^{th}$ SOM layer. This procedure is illustrated in Fig. 5.13, where the maximum number of children nodes of a node at the $(k\text{-}1)^{th}$ level is 6. Therefore the number of the 1-D SOM neurons used for the training of output positions at the kth SOM layer is 6. After the completion of training of the 1-D SOM, the 2-D weight vectors of all the 6 neurons are marked in circles with index labels as shown in Fig. 5.13. The neighboring neurons are connected with a solid line. Consider three nodes at the kth level: *A*, *B* and *C*. p_A, p_B and p_c are their corresponding output positions on the kth SOM layer. The three nodes are the children nodes of a parent node at the (*k*-1) level. The 2-D position vectors p_A, p_B and p_c are marked in cross symbols as shown in Fig. 5.13. The ordering algorithm just assigns p_A to its nearest neuron, i.e., neuron index 3. Then p_B and p_c is assigned to neuron index 2, neuron index 6 respectively. Therefore the SOM input representation $[p_1, p_2, p_3, p_4, p_5, p_6]$ of their parent node at the (*k*-1) level is $[\mathbf{0}, p_B, p_A, \mathbf{0}, \mathbf{0}, p_c]$, where $\mathbf{0}$ is [0,0]. This ordering makes the later similarity measurement more reasonable.

5.6.2. *MLSOM Training*

Assume that each node at the kth level of trees has a n_k dimensional feature vector. The maximum of children nodes of a node at the kth level are c_k. The maximum levels of trees and maximum layers of MLSOM are all *L*. The input data for the kth SOM layer are all $n_k + 2c_k$ dimensional vectors. There are m_k neurons at the kth SOM layer. The weights of neurons at the kth SOM layer are also $n_k + 2c_k$ dimensional vectors. The learning steps of the MLSOM for a tree-structured data are described as follows.

1. Set the current level *k* of trees and the current layer *k* of an MLSOM to be *L* (bottom level and bottom layer).

2. Normalize the 2-D positions of neurons in each SOM layer in a range of $((0,1],(0,1])$.
3. Set iteration t to be 0. Collect all the nodes at the kth level nodes of trees to be the training data for the kth SOM layer. The ordered 2-D SOM output positions of children nodes of the nodes at the kth level are filled in the SOM input representation at the kth SOM layer. Therefore the $n_k + 2c_k$ dimensional vectors are generated for the inputs of the kth SOM layer. Normalize the values in each dimension of them in a range of [0,1]. Randomly initialize the $n_k + 2c_k$ dimensional vectors for the m_k weights at the kth SOM layer.
4. Randomly select a vector x from the $n_k + 2c_k$ dimensional vectors for the inputs of the kth SOM layer.
5. Find the winner neuron a at the kth SOM layer:
$$a = \arg\max_i \left\| S(x, w_i^k) \right\|, \qquad i = 1, \ldots, m_k \tag{5.34}$$
where $S(x,\ w_i^k)$ is the similarity measurement of x and w_i^k, w_i^k is the weight vector of the ith neuron at the kth SOM layer. The similarity measurement of x and w_i^k is defined as follows.
$$S(x, w_i^k) = \frac{\lambda}{n_k} \sum_{j=1}^{n_k} \left\{ 1 - abs\left(x_j - w_{ij}^k \right) \right\} + \frac{1-\lambda}{\sum_{j=1}^{c_k} \delta(x_{2j+n_k-1}, x_{2j+n_k})}$$
$$\sum_{j=1}^{c_k} \left\{ \delta(x_{2j+n_k-1}, x_{2j+n_k}) \left[1 - \sqrt{\left(x_{2j+n_k-1} - w_{i(2j+n_k-1)}^k \right)^2 + \left(x_{2j+n_k} - w_{i(2j+n_k)}^k \right)^2} \right] \right\} \tag{5.35}$$
where x_j is the jth value of x, w_{ij}^k is the jth value of w_i^k, λ is a weighting parameter, $\delta(x, y)$ is a function such that
$$\delta(x,y) = \begin{cases} = 1, & \text{if } x \neq 0 \text{ and } y \neq 0 \\ = 0, & \text{otherwise} \end{cases}$$
The first term in (5.35) considers the features of the current node whilst the second one considers the compressed features of its children nodes. The weighting parameter λ determines the

emphasis on global features, which appears at root nodes or higher level, or local feature that appears at lower part of the tree. The choice of λ is problem-dependant. An equal weighting (λ=1) is used in general unless specified.

6. Update the weights of neurons at the kth SOM layer by
$$w_i^k = w_i^k + \eta(t)h_{ia}(t)\left(x - w_i^k\right), i = 1,\ldots,m_k \qquad (5.36)$$
where $\eta(t)$ is the learning rate at iteration t, $h_{ia}(t)$ is the neighborhood function around the winner node a.
7. Increase the iteration by $t = t + 1$.
8. When the maximum number of iterations is reached, the training at the kth SOM layer stops. Otherwise go to step 4.
9. If the ordering of children nodes is not predefined, the 2-D SOM output positions of nodes at the kth level are ordered by an ordering algorithm, otherwise, go to step 10.
10. When k is equal to 1 (top level of trees or top SOM layer), the training stops. Otherwise, $k = k - 1$, go to step 3.

From the above training steps, the second parts of the SOM input representation are fixed before training. The nodes at each level are fed into the corresponding SOM layer for training. Therefore the computational complexity of the MLSOM for one epoch is $O\left(\sum_{k=1}^{L} N_k m_k \left(n_k + 2c_k\right)\right)$, where N_k is the number of nodes at the kth level of trees, m_k is the number of neurons at the kth SOM layer, n_k is the number of input features of a node at the kth level, and c_k is the maximum number of children nodes of all trees at the kth level.

5.6.3. MLSOM Visualization and Classification

The top SOM layer in an MLSOM plays an important role for visualization of the tree-structured data. After the completion of the MLSOM training, the top SOM layer can be used for data visualization. As each root node represents each tree structure, the tree-structured data can be visualized by their associated neurons. Some useful information, such as clustering tendency, can be further detected from the top SOM layer.

The MLSOM can be further used for classification. A neuron at the top SOM layer can be labeled with a class that most associated tree-structured data belong to. In the case of tie, the class is chosen randomly. If a neuron is not assigned any datum, we just search the nearest neurons and the neuron is labeled with the class that the most structured data associated with the nearest neurons belong to. Thus the SOM map is labeled by assigning data classes to all neurons. The labeled SOM map, called "class map", can be used for classification. When a testing tree-structured datum is presented to the MLSOM, a best-matching neuron at the top layer can be found according to Eq. (5.34). As the best-matching neuron is already labeled with a class, the testing datum is classified to that class.

We used a real data set consisting of 480 real flower images to demonstrate the processing of a tree-structured data using the MLSOM. There are totally 12 different species of flowers. Each species has 40 flower images. All the flower images are divided into training and testing sets. The total number of flower images used for training is 240 and that for testing is 240. Fig. 5.14 shows a tree-structured data extracted from a flower image. A flower image is represented by a three-level tree. The root node of a tree represents the whole image. The 2nd level nodes of a tree represent local regions such as background and flowers. The 3rd level nodes of a tree represent more specific local regions such as each individual flower. Different types of features are used to describe nodes at different levels as shown in the figure. An MLSOM with 3 layers (40×40-48×48-56×56) is used to process the flower images. After the training is completed, the visualization on the top layer is shown in Fig. 5.15. The 12 species of flowers from the training set are shown on the map in different symbols. The map indicates different flowers form different clusters on the output map. Using the above-mentioned "class map" procedure, a classification performance of 100% and 93% are achieved on the training set and the testing set respectively.

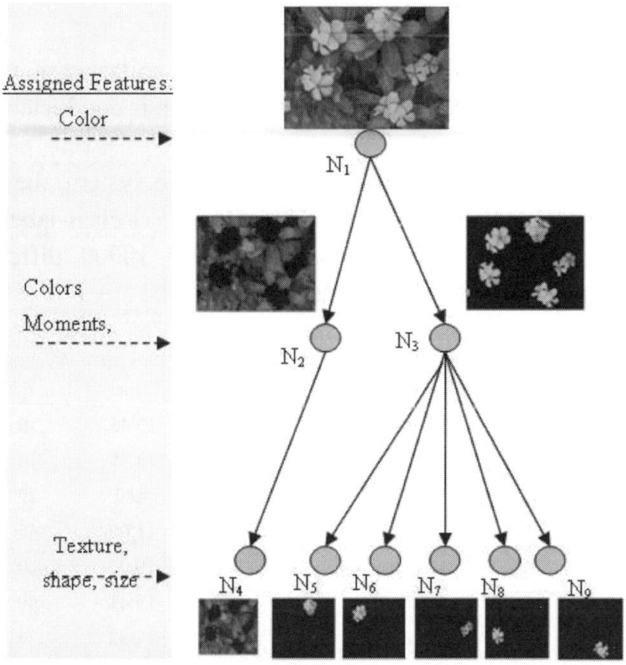

Figure 5.14. Illustration of the tree-structured flower image representation. Different types of features are assigned to different levels of nodes

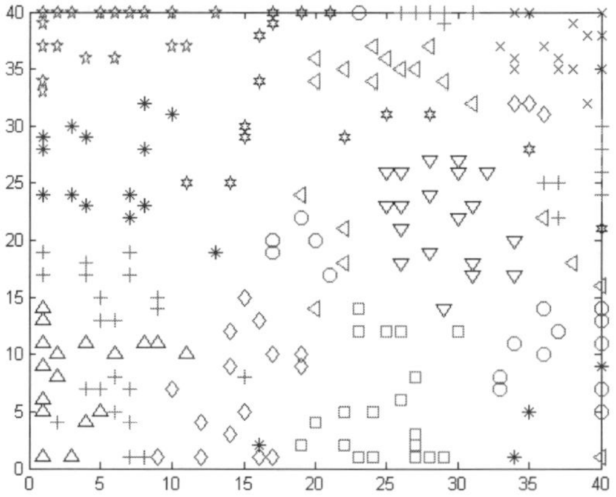

Figure 5.15. Visualization of 12 different flower species on the top layer of an MLSOM

Exercises

Q5.1. An SOM is used to learn a dataset with no target output. What does the SOM learn and how visualization is conducted?

Q5.2. In an electric power plant fault diagnostic system, the following table shows 9 data points of a 8-attribute + 1 class-label data file. The whole data file consists of over 10000 different fault instances collected in the past.

Circuit ID	Weather type	Phases	Loading type	Season	plant	Cust off time	Duration in min	Fault Causes
1	o	R	5 (High)	S	c4	13:45	10	F1
1	o	R	5 (High)	S	c4	14:45	10	F2
2	o	R	3 (Medium)	A	c4	9:40	20	F3
2	1	Y	5 (High)	A	x	11:00	45	F4
3	10	Y	5 (High)	A	x	20:00	20	F1
3	2	Y	1 (Low)	W	c4	19:43	60	F4
3	3	R	4 (Quite high)	W	c4	17:32	20	F3
5	3	R	3 (Medium)	W	c4	16:44	13	F3
6	3	R	4 (Quite high)	W	c4	11:33	7	F2

Which of the above attributes should not be used in the SOM training for studying plant fault diagnosis?

Q5.3. In Q 5.2, there are only 6 types of fault, i.e., ice, animal, lightning, human, tree, and deterioration denoted F1, F2, F3, F4, F5 and F6 under the attribute of fault causes. These six types of fault are found to be nicely discriminated by an SOM using 10x10 neurons. Sketch a typical SOM map to show the possible result. In this case, what is the factor affecting the output visualization of the SOM map?

Q5.4. The weight adaptation of the SOM learning rule uses a simple compctitivc lcarning mcthod with ncighborhood function bcing set to h=1 for the winner neuron, and =0.5 for all other neurons. Also, assume η=0.5. In a data set, the current input data x1=[2, 3], and the following input data are [1, 4], and [3, 1], the initial weights of three neurons are w1=[-1, 2], w2=[2, 0] and w3=[1, 1]. Show how the weights are updated for only one number of iteration.

Chapter 6

Classification and Feature Selection

6.1. Introduction

To mimic human's ability in discriminating various objects using mathematical models and computer has always been an intriguing issue for human beings. Among many other AI, or pattern recognition techniques, the use of neural networks to perform discrimination task has proved to be appealing because of its supervised and unsupervised learning ability. Neural classifiers have shown many important applications since computers have become immensely powerful compared to a decade ago. This enables us to conduct a lot of complicated classification or recognition tasks that were difficult or even impossible to be handled in the past decades. We are now able to design classifiers capable of discriminating between members and non-members of a given class as a result.

When we use neural networks to act as a classifier, we are simply finding a network to represent a non-linear discriminant function. As a result, the trained network will deliver a classification result when new input vectors are presented to it. On the other hand, we can view that the network is used to model the posterior probability of class membership. Posterior probability, a quantity widely used in statistical analysis, enables us to make a near-optimal decision on the class membership of a given data. Classifier is then considered as a machine whose decision on class membership is made according to the probability distribution of the variables.

For a given feature vector x, classifying its class membership requires the knowledge of its posterior probability. For a given feature vector x, the posterior probability, $P(C_a|X)$, tells us the probability of the instances belonging to class C_a. Maximizing the probability of classification can be obtained by finding the class C_a having the largest posterior probability. Conversely speaking, classifying newly given feature vector x as certain class having the largest posterior probability is equivalent to minimizing the probability of misclassification.

In practical applications, a classifier can be designed by finding appropriate discriminant functions, which are usually in forms of parametrized functions. In most cases, the values of the parameters are determined from a set of training data together with a learning algorithm. For example, we consider a simple problem of distinguishing men from women. Certain feature vector X_i, of a given human face must be extracted and analyzed. We might, for instance, analyze the pixel-based representation of face images. We might expect that the feature vector X_i, transformed from pixel-based codes, of women face images will be different from the men in certain extent. Apparently, for certain given feature vectors, the probability of misclassifying the face image will be low, which means that we will be very certain that it is either a woman or a man. On the other hand, it is perceivable there are many cases that the class membership overlaps with feature vectors, which means that there is a high chance of misclassifying. A classifier can then be seen as a system which consists of a set of input vectors mapping to an output variable y, which in this case representing the label of whether it is a man or a woman. The mapping is done using certain mathematical modeling. For instance, the mathematical model can be in the form $y_i = g(X_i; W)$, where the model can be considered as a neural network, and W are the weights of a neural network. By optimizing the parameters W, we will be able to minimize the probability of misclassifying a given facial image. Perceptron, regarded as one of the earliest form of neural classifier, is a well-known linear discriminat developed in the early sixties for recognizing simple characters, although it was later proved to be unable to handle a linear non-separable problem.

In Chapter 1, we have shown the principle of its operation and how its weights are determined.

There are several types of widely used classifiers including K-nearest neighbour, multi-layer Perceptron (MLP), Radial Basis Function network, Support vector machines (SVM), Bayesian classifiers, and Decision Tree. They are either in the category of statistical classifier or machine learning classifier. Some of these have been thoroughly described in the previous chapters. Classifiers have been successfully applied to a wide range of applications including handwriting recognition, finger print recognition, speaker identification, and text categorization, faults diagnostic, etc. Most recently it has been extended to the application of handling the emerging micro-biology problems. The advent of complementary DNA (cDNA) microarrays technology enables computer scientists to measure the expression levels of thousands of genes in a single experiment. cDNA microarrays technology allows detection of the expression levels of thousands of genes at a time. Bioinformatic researchers, biologists and medicine researchers are now able to compare different ribosomal proteins or genes coding for all sorts of medical purposes. This provides a new way to understand and investigate the differences in the gene expression patterns.

In microarrays cDNA classification, neural classifiers are used for classifying whether a given patient's data is cancer positive or not. Prior using a classifier, advanced features selection technique is used to extract the informative genes among the thousands genes because a microarrays dataset usually consists of thousands of features and classifiers usually find handling a huge feature set difficult. Thus in the later section of this chapter, some classification and feature selection methods will be discussed.

Tables 6.1 show interesting results obtained from a 10 fold test on different diseases and classifiers. The classifiers studied include *k*-NN, MLP, SVM, and Decision Tree. In the tests, only the most prominent 50 features are selected from a feature selection scheme for classification. The numbers of data points are small so they may experience the small data point problem. This can be clearly seen in the Colon cancer case which consists of 62 data points only. Apparently, the classification accuracy is significantly lower than the Lung Cancer dataset which consists of 181 data points. In this type of bioinformatics dataset, it is generally noticed that SVM can deliver very promising performance in terms of classification accuracy. SVM and MLP are found to be relatively more robust to noise corrupted data, while KNN and decision Tree can provide very stable performance under noise free environment.

Table 6.1 (a) Details of the four investigated cancer diseases

Dataset Name	Number of data	Dimensions	Number of selected features
Colon Cancer	62	2000	50
Prostate Cancer	102	12600	50
Lung Cancer	181	12533	50
ALL-AML Leukemia	72	7129	50

Table 6.1 (b) Classification accuracy of Colon Tumor data (10 fold test)

	KNN	MLP	SVM	Decision tree
Mean value	0.693	0.7234	0.7412	0.6962
Std V value	0.0089	0.0081	0.0098	0.0057

Table 6.1 (c) Classification accuracy of Prostate Cancer data (10 fold test)

	KNN	MLP	SVM	Decision tree
Mean value	0.834	0.8684	0.8725	0.8184
Std V value	0.0046	0.0036	0.0049	0.0086

Table 6.1 (d) Classification accuracy of Lung Cancer data (10 fold test)

	KNN	MLP	SVM	Decision tree
Mean value	0.9831	0.9625	0.9874	0.9547
Std V value	0.00078707	0.0025	0.0014	0.0013

Table 6.1 (e) Classification accuracy of ALL-AML_Leukemia data (10 fold test)

	KNN	MLP	SVM	Decision tree
Mean value	0.824	0.8541	0.8902	0.8981
Std V value	0.0074	0.0045	0.0047	0.0053

The above results indicate that different classifiers perform slightly different on the same cancer classification. The factors affecting their classification accuracy are very complicated including the noise level of the data, the level of redundancy data, the dimensionality of the problem, and number of data sample etc. But this is usually perceived as problem dependent.

6.2. Support Vector Machines (SVM)

Support vector machines (SVMs) are a new type of supervised learning methodology nicely developed by Vapnik and his associates. It is an elegant and powerful approach for performing classification based on the principle of structural risk minimization, for instance, minimizing the summation of empirical risk and the bound of risk confidence. It is worth pointing out that most classifier like multi-layer Perceptron network can perform classification, but they are prone to the risk of finding trivial solutions that overfitting the data. Its learning algorithms do not guarantee to minimize the margin errors over all hyperplanes. The advantages of SVMs over traditional multi-layer Perceptron networks are their better generalization ability and global minimization.

A typical classification problem is the separation of positive members from negative members. Generally, we are required to build a conventional classifier separating the positive members from negative members. If the data points in the training set are vectors of m numbers, the development of such a classifier is to find a hyperplane that is able to separate the members. In most real-world problems, the problems are usually not ideal and they involve non-separable data. This means that there may not be a hyperplane allowing us to separate all the positive members from the negative members. SVMs map a given training set into a possibly high-dimensional feature space in an attempt to determine a hyperplane that separates the positive members from the negative members. Much of the beauty of SVM's comes from its elegantly defined criterion for selecting a separating plane because there may usually be many high-dimensional candidate planes available for doing a similar job. SVM is able to select the plane that maintains a maximum margin in the training set. This is important because statistical learning theory suggests that the choice of the maximum margin hyperplane will lead to maximal generalization when the result is used for predicting classification of unseen data. The theory enables us to find a separating hyperplane in the feature space simply by defining a Kernel function.

Let the training data be $\{x_i\}, i = 1,...,N$. The corresponding class labels are $\{y_i \mid y_i \in \{-1,1\}\}$. If a hyperplane separates the two classes successfully, the data point x lying on the hyperplane satisfies $w \cdot x + b = 0$. Let $d_{\min}$ the minimum distance from the hyperplane to a datum with label "1" or "-1". In SVM, $d_{\min}$ is selected as 1. Then the training data satisfy

$$w \cdot x_i + b \geq 1, \quad \text{for } y_i = 1 \tag{6.1}$$

$$w \cdot x_i + b \leq 1, \quad \text{for } y_i = -1 \tag{6.2}$$

Thus Eqs. (6.1) and (6.2) with equality define two parallel hyperplanes H1 and H2. The above inequalities can be combined into one inequality:

$$y_i(w \cdot x_i + b) - 1 \geq 0, \forall i \tag{6.3}$$

The distance between H1 and H2 is $2/\|w\|$, which is also called margin. SVM tries to maximize the margin to deliver the good classification performance, under the constraint of Eq. (6.3). This can be expressed by a Lagrangian:

$$\text{minimize} \quad L_p = \frac{1}{2}\|w\|^2 - \sum_{i=1}^{N} a_i y_i (w \cdot x_i + b) + \sum_{i=1}^{N} a_i, \ (a_i > 0) \tag{6.4}$$

The above problem can be converted to a dual one and the solution of classification can be obtained using an optimization method. For a non-separable case, slack variables can be introduced in Eqs. (6.3) and (6.4). Despite the excellent classification ability of SVMs, they are unable to provide a visualized classification result. Class boundary between the two classes cannot be visualized for users. The process may appear to be obtained from a mathematical tool-box only. In some cases, this may be rather restrictive when boundary information is useful for enhancing the classification results. The boundary information may provide useful information on the relative distance of the instances during classification. The next section will then describe a new type of classification tool called SVM visualization (SVMV).

6.2.1. Support Vector Machine Visualization (SVMV)

SVMV is a useful tool to visualize the SVM classification results. SVMV is developed by exploiting the advantages of both SVM and SOM. SVMV gives additional information about the classification boundary. That is, data close to the class boundary can be classified directly from visualization. The distance between a datum and classification boundary can be directly visualized in a low-dimensional space. As a result, we are able to minimize the possibility of mis-classification of data when they appear to be too close to the classification boundary. The whole classification results obtained by SVM can be clearly elaborated through visual clarification.

In general, there are two main types of SVMs models for binary classification: C-SVMs and v-SVMs. But in terms of classification, v-SVMs are better than C-SVMs because the parameter $v \in (0,1)$ in v-SVMs has salient physical meaning and is thus used in this section to illustrate the theory of the SVMV. Suppose an empirical dataset $(x_1, y_1), \ldots, (x_n, y_n)$, where $x_i \in \Re^d$ is a d-dimensional input vector, $y_i \in \{\pm 1\}$ is the binary class label, n is the number of input data. As illustrated in Fig. 6.1, v-SVMs try to separate two classes using an optimal hyperplane that maximizes the margin of separation. It solves the following primal problem:

$$\begin{cases} \min & \phi(W) = \frac{1}{2}\|W\|^2 - v\rho + \frac{1}{n}(\sum_{i=1}^{n} \xi_i) \\ \text{s.t.} & y_i[(W \cdot \varphi(x_i)) + b_0] \geq \rho - \xi_i, \\ & v > 0, \xi_i \geq 0, \rho \geq 0, i = 1, 2, \ldots, n \end{cases} \tag{6.5}$$

where φ is a mapping function from input space to high-dimensional feature space, $W \in \Re^d$ is a vector perpendicular to the optimal hyperplane in feature space, $b_0 \in \Re^d$ is a bias vector, ξ_i is a slack variable for allowing classification errors, $\rho \in \Re^d$ is another bias vector, and v is a parameter discussed before. Instead of solving problem Eq. (6.5) directly, v-SVMs solve its dual problem as follows:

$$\begin{cases} \max \; Q(\alpha) = -\frac{1}{2}\sum_{i,j=1}^{n} \alpha_i \alpha_j y_i y_j \varphi(x_i)^T \varphi(x_j) \\ \text{s.t.} \quad 0 \le \alpha_i \le \frac{1}{n}, \quad \sum_{i=1}^{n} \alpha_i y_i = 0, \\ \quad \sum_{i=1}^{n} \alpha_i \ge v, i = 1, 2, \ldots, n, \end{cases} \tag{6.6}$$

where α_i is a Lagrange coefficient. The solution of α_i in Eq. (6.6) can be used to compute w, b_0, and ρ in (1) [20]. To avoid computing the dot product in the high dimensional feature space, SVMs use kernel functions. After the completion of optimizing Eq. (6.6), the data points with $0 < \alpha_i < 1/n$ are called support vectors (SVs).

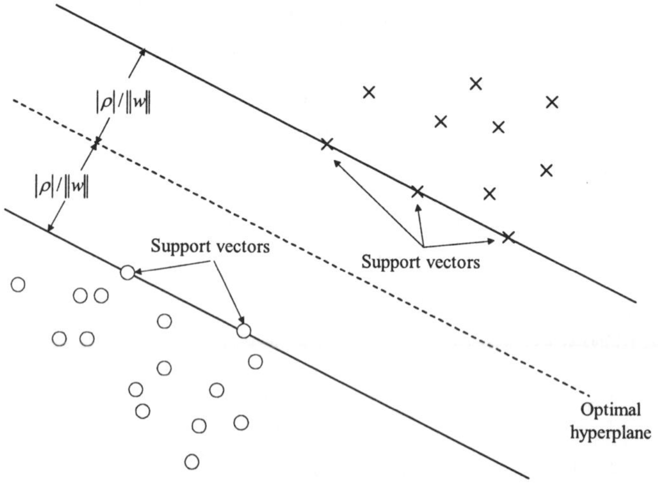

Figure 6.1. Illustration of the separation of two classes by v-SVM

The decision function for the binary classification of the data is

$$\mathrm{sign}(W \cdot \varphi(x) + b_0) = \mathrm{sign}(\sum_{i=1}^{n} \alpha_i y_i K(x_i, x) + b_0) \tag{6.7}$$

The bias function for a new input z is

$$f(z) = \left(\sum_{i=1}^{n} \alpha_i y_i K(x_i, z) + b_0 \right) \tag{6.8}$$

The distance between z and the optimal hyperplane (classification boundary) in feature space is

$$|f(z)| / \|w\| \tag{6.9}$$

For the SVs with $0 < \alpha_i < 1/n$, the value of Eq. (6.8) is ρ or $-\rho$, and that of Eq. (6.5) is $\rho / \|w\|$. For the data points lying in the classification boundary, the value of Eq. (6.8) or Eq. (6.9) should be zero. In SVMV, SOM is firstly trained. The SVMs are then applied to the data to find the optimal solution of classification. Using the result from SVMs, the distances (or bias) between feature vectors of neurons of the SOM and optimal hyperplane of SVM in high-dimensional feature space can be obtained. The classification boundary can be found and visualized in the neurons with zero distance (bias) to optimal hyperplane. The detailed procedure of the SVMV algorithm is described as follows:

Step 1. First, an SOM algorithm is trained to obtain the topology-preserved weights $\{w_i\}$, i=1,..., m.

Step 2. Then ν-SVMs are used to obtain the optimal value of α_i, $i = 1,...,n$, according to Eq. (6.6).

Step 3. Interpolation is performed between the neighboring pairs of weights $\{w_i\}$ in SOM and therefore extended weight vectors $\{w_i'\}$ are formed to obtain a more precise map. The detailed interpolation will be explained in the next subsection.

Step 4. After that, the bias function $f(w_i')$ on the extended weight vectors $\{w_i'\}$ is computed according to Eq. (6.10):

$$f(w_i') = \sum_{i=1}^{n} \alpha_i y_i K(x_i, w_i') + b_0 \tag{6.10}$$

Step 5. Finally a 2-D map of SOM is colored according to the values of $\{f(w_i')\}$. The neurons with grey color, whose bias $f(w_i')$ is larger than 0, belong to class "1". On the contrary, the neurons with white color, whose bias is less than 0, belong to class "–1". Obviously, the bias $f(w_i')$ of the neurons on the boundary is equal to zero. So the classification boundary can be detected by $f(w_i') = 0$.

Due to the topology preservation of SOM, if a datum in the input space is close to classification boundary, they are also close to each other in the reduced low-dimensional space. The distance between data and classification boundary can be displayed on the reduced low-dimensional space. In Fig. 6.2, they show some of typical examples of using SVMV for classification of leukemia dataset which consists of 72 leukemia samples with 7129 genes. The following results show the training data contain 27 acute lymphoblastic leukemia (ALL) samples and 11 acute myeloid leukemia (AML) samples. The testing data contain 20 ALL samples and 14 AML samples. AML samples are labeled with class "+1" while ALL samples are labeled with class "1". The kernel width σ is selected as 1 and 10. The AML and ALL samples are visualized by cross and dot symbols, respectively. For the training data, the data with the same class are located in the region of the same color. This indicates the training data are well classified with 100% classification accuracy. For the testing data shown in Fig. 6.2(b), there is only one neuron out of the 4 AML samples located in the white region which represents ALL samples. It means a classification error of 11.8% for the testing data. SVM with a small RBF kernel width means that the classification boundary is more complicated. Although the classification can be very good for training data, it has poor generalization for classifying testing data. In Fig. 6.2 (c) and (d), when the RBF kernel width σ is selected as 10, the generalization of SVM is improved. The classification accuracy is 100% for both the training and testing data. In this section, the SVMV method is a hybrid approach of both SVM and SOM. It achieves good visualization effect with classification boundary. Unlike SVM that cannot explain the whole classification mechanism, the SVMV provides 2-D visualization on complicated classification results.

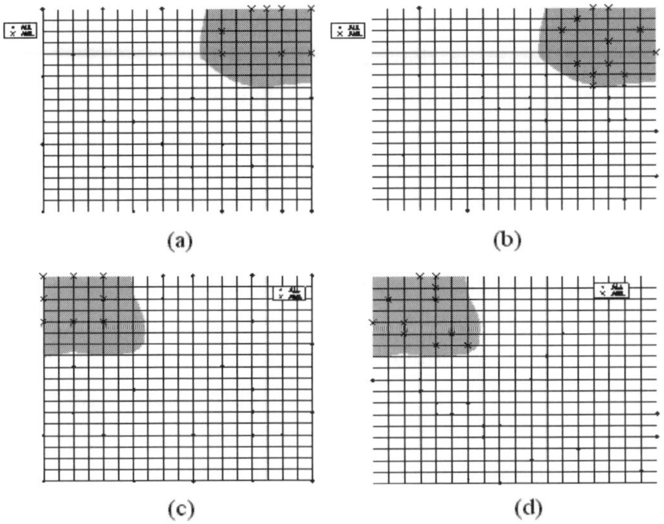

Figure 6.2. Visualization results on the leukemia dataset by SVMV (a) (σ=1) Visualization on the training data (b) (σ=1) Visualization on the testing data (c) (σ=10) Visualization on the training data (d) (σ=10) Visualization on the testing data

6.3. Cost Function

In previous chapters, we have shown the use of mean squared error (MSE), and higher-order cumulants cost functions on different learning algorithms. The choice of cost function is dependent upon the applications. We have also shown that a careful design of cost function can enhance the generalization, and convergence rate of the network performance. For instance, for general regression problems of which the objective is to model the relationship between the input variables and output variables. The use of MSE or MSE-FOC cost function will have a positive effect of enhancing the learning performance. For classification problem, the objective is largely different. A classifier is to model the posterior probability of class membership with input variables. When MSE cost function is used, there are differences between the objectives of training a neural classifier and delivering good classification performance. By minimizing the MSE cost function, the training algorithms are aimed to make the output of a network approximates the discrete values, such as 0 and 1, as close as possible. But a classification task is generally made on the basis of hard decision rules. That is, we classify a given observation to the class having the maximum output in order to minimize the classification error probability. In this sense, the requirements on the output of a classifier can be relaxed. It is largely satisfactory that the output of a classifier can correctly distinguish different class patterns, although the value of output may not be able to approximate the designated value (1 or 0) closely. In a training process of using MSE cost function, the patterns that have been correctly recognized still have an unnecessary contribution to the subsequent training process. This may reduce the convergence rate and cause overfitting.

When we are designing classifiers, our aim is clearly to enhance the classification ability, the minimum classification error function (MCE), which was proposed for achieving minimum classification error, is a good choice over the conventional MSE. The concept of MCE cost function is to find a set of classifier parameter together with their associated decision rule such that the probability of misclassifying any given observations can be minimized.

Although the MCE function enables the neural classifier to be directly constructed to minimize the classification error, there is certain difficulty in implementing MCE based classifiers. The selection of a smoothness parameter ξ of MCE has a marked effect on training a classifier. A large ξ results a rather rugged error surface in which a lot of local minima may cause sub-optimal results. On the other hand, when ξ is set to a small value, the error surface of the MCE becomes smoother. Dynamically changing of ξ is an option to improve the classification performance and the convergence rate. It is an algorithmic approach to adjust the smoothness parameter ξ when the training proceeds but requires more implementation effort. In the following sections, we discuss the MCE and a hybrid MSE-MCE cost functions for classification.

6.3.1. MSE and MCE Cost Functions

Assume that patterns in dataset $X = (x_1, x_2, ..., x_N)'$ are fallen into J categories. We also assume a classifier has J output units, for instance, one output unit corresponds to one class. Below, for convenience, for a data pattern x_i, let $t_i = (t_{i1}, t_{i2}, ..., t_{iJ})$ and $y_i = (y_{i1}, y_{i2}, y_{i3}, ..., y_{iJ})$ be the output target and the actual output of a classifier, respectively. For a pattern (say, x_i) belonging to the class $cl_j (j = 1, 2, \cdots, J)$, it is defined that

$$t_{ik} = \begin{cases} 1, & k = j \\ 0, & k \neq j \end{cases}$$

Mean square error function, E_{mse}, commonly used for training classifiers, can be expressed as

$$E_{mse} = \frac{1}{2N} \sum_{i=1}^{N} \sum_{j=1}^{J} (t_{ij} - y_{ij})^2 \tag{6.11}$$

In Eq. (6.11), the classification decision rules are not directly reflected. Through minimizing Eq. (6.11), the output of classifier approximates the target (1 or 0) as close as possible. A hard decision rule is directly adopted in the MCE function [11, 12]. In the MCE function, $y_i = (y_{i1}, y_{i2}, y_{i3}, ..., y_{iJ})$ are the outputs of a set of the discriminant functions. Generally, the class is determined as

$$x \in cls_k \quad \text{if} \quad y_k(x) = \arg\max_{1 \le j \le J} y_j(x)$$

Based on this type of decision rule, for x belongs to the class cl_m, recognizing x correctly requires y_m to be larger than other y. Also, the larger the difference between y_m and other y is, the smaller the misclassification risk will be. Thus, the difference between y_m and the largest y can be used to measure the classification ability of the built discriminant function set. This is the rationale behind the misclassification measure Eq. (6.12) defined in [11].

$$d^{(m)}(x,\Lambda) = -y_m(x,\Lambda) + \left[\frac{1}{M-1}\sum_{j,j\neq m} y_j(x,\Lambda)^p\right]^{\frac{1}{p}} \tag{6.12}$$

where p is a positive constant, and Λ be the parameter set of the discriminant functions. A small value of $d^{(m)}$ indicates a low misclassification probability. With different p, one can take all the potential class into consideration to a different degree. When p is large enough, we have

$$\arg\max_{j,j\neq m}(y_j(x,\Lambda)) \approx \left[\frac{1}{M-1}\sum_{j,j\neq m} y_j(x,\Lambda)^p\right]^{\frac{1}{p}}$$

That is, for a large p, Eq. (6.12) is able to directly reflect the hard decision rule. We can set that $p = 10$. For convenience, we rewrite Eq. (6.12) in a form of output vector y_i and target vector t_i

$$d^{(m)}(x_i,\Lambda) = -t_i \times y_i' + \left[\frac{1}{M-1}(1-t_i)\times(y_i^p)'\right]^{\frac{1}{p}} \tag{6.13}$$

where $(\bullet)'$ means the transpose of vector $\bullet$. In Eq. (6.12) or Eq. (6.13), $d^{(m)} < 0$ indicates a correct class determination, whereas $d^{(m)} \geq 0$ results in a misclassification determination. With $d^{(m)}$, the misclassification error function (MCE) is

$$\ell(x_i,\Lambda) = \frac{1}{1+e^{-\xi(d^{(m)}(x_i,\Lambda))}} \tag{6.14}$$

where $\xi > 0$, and $x_i \in cls_m$. Obviously, Eq. (6.14) increases with the decreasing of $d^{(m)}$. Optimizing Eq. (6.14) can directly maximize the classification ability of the constructed model. Eq. (6.14) is a smooth zero-one function and can be optimized using gradient decent type algorithms. The MCE on the whole dataset $X = \{x_1, x_2, \ldots, x_N\}$ is the mean of MCE results on all patterns, for instance,

$$E_{mce} = \frac{1}{N}\sum_{i=1}^{N}\ell(x_i,\Lambda) \tag{6.15}$$

The MCE Eq. (6.15) evaluates the classification ability of a classifier in a direct way. This is a clear advantage of enhancing the classification performance.

6.3.2. Hybrid MCE-MSE Cost Function

First we would like to focus on analyzing the MCE function before we move to the hybrid MCE-MSE cost function. In a batch version of gradient descent learning, the direction of updating is the average of negative gradient of all patterns. It is expected that the patterns far from being correctly classified should impose more effect on the parameters adjustment, while the patterns close to the target should have less effect. In this sense, the MSE function satisfies the requirement well. But, the MCE function Eq. (6.16) is unable to satisfying the above requirement because $|E_{MCE}'|$ decreases with the increase of d_m for misclassified patterns (i.e., $d_m > 0$). This shows that the data patterns that are far from being correctly classified may have small or even negligible effect on adjusting the parameters.

For example, for a data pattern x, the real output is $y = (y_1, y_2)$ and the target output is $t = (1,0)$. We have the 2-dimensional MCE function

$$E_{mce} = \frac{1}{1 + e^{-\xi(-y_1 + y_2)}} \tag{6.16}$$

Fig. 6.3 illustrates the relationship between $y = (y_1, y_2)$ and the derivative $\partial E_{mce}/\partial y_1$. Table 6.1 lists several values of $\partial E_{mce}/\partial y_1$, and $\partial E_{mse}/\partial y_1$, and their corresponding outputs. Take point 1, point 2 and point 3 as examples. The situation of point 1 certainly satisfied the above requirement because its output is close to the target and $\partial E_{mce}/\partial y_1$ is very small. But for point 2 and 3, it is desirable to have a larger value of $\partial E_{mce}/\partial y_1$ to minimize the classification error. The $\partial E_{mce}/\partial y_1$ values of these points, however, are small. This illustrates that using gradient decent type training methods to optimize E_{mce} may not assure a satisfactory convergence rate and the subsequent classification results. This shortcoming can be addressed by dynamically controlling the smoothness parameter ξ, but this approach is rather problem dependent and may not be straightforward for implementation.

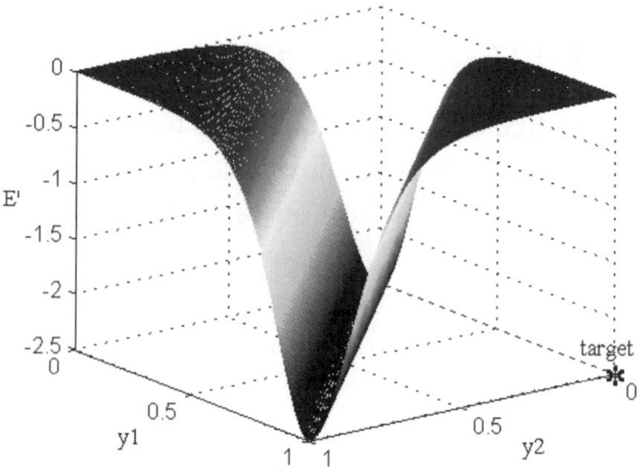

Figure 6.3. It shows the relationship between $\partial E_{mce}/\partial y_1$ and the classifier output $y = (y_1, y_2)$. Supporting that, for this 2-class, the target for an input data pattern is $t = (1,0)$. Its MCE function is $E_{mce} = [1 + e^{-\xi(-y_1+y_2)}]^{-1}$

Table 6.2 Gradient value of MCE function and MSE function

Target [y1, y2] = [1 0]			
Point No.	Output [y1, y2]	Gradient of MCE $\left[\frac{\partial E_{mce}}{\partial y_1}, \frac{\partial E_{mce}}{\partial y_2}\right]$	Gradient of MSE $\left[\frac{\partial E_{mse}}{\partial y_1}, \frac{\partial E_{mse}}{\partial y_2}\right]$
1	[0.9 0.1]	[0.0031 -0.0031]	[0.1000 -0.1000]
2	[0.5 0.9]	[0.3133 -0.3133]	[0.5000 -0.9000]
3	[0.1 0.9]	[0.0061 -0.0061]	[0.9000 -0.9000]

We can consider a hybrid cost function

$$E = \frac{1}{2}(E_{mse} + E_{mce}) \tag{6.17}$$

The MCE-MSE cost function encompasses the properties of both MCE and MSE. The MSE part of this cost function enables the patterns with large classification errors to be taken into account irrespective of the value of ξ in the MCE. The shortcoming of MCE is compensated in a more simple way. The MCE-MSE function maintains the balanced property of MCE and MSE functions and provides a balanced performance on convergence rate and classification results. Similar to the MCE function, selecting different value of smoothness parameter ξ affects the performance of the MCE-MSE based learning algorithm, and usually ξ can be fixed as 1 for most problems.

We use RBF classifier solving a two-spiral classification problem as example. The training processes are stopped when the classification error has not further decreased for 100 training epochs continuously. The averaged results of 10 trials are listed in Table 6.2 where ξ of the MCE function is 1. It shows that both the number of epochs and running time required by the MCE cost function are significantly larger compared with the proposed MCE-MSE based cost function.

Table 6.3 Comparisons of different cost functions

Cost function	Number of epochs	Running time (sec)	Classification error (%)
MCE-MSE	5380	267	2.0
MSE	6040	383	12.6
MCE	15,340	1.31×10^3	1.9

Typical convergence curves are shown in Fig. 6.4. Due to the complex boundary of this problem, the training process generally begins with 50% classification error. That is, a major part of patterns are far away from being correctly recognized. To swiftly reduce the classification error requires weighing the misclassified patterns more than the correctly-classified patterns. MSE function is able to meet this requirement, whereas MCE function has difficulties to satisfy this condition. Thus, the convergence rate of MSE and the MCE-MSE based gradient algorithms are much faster than the MCE based approach. In terms of classification results, the hybrid MCE-MSE based algorithm shows promising performance compared with other approaches.

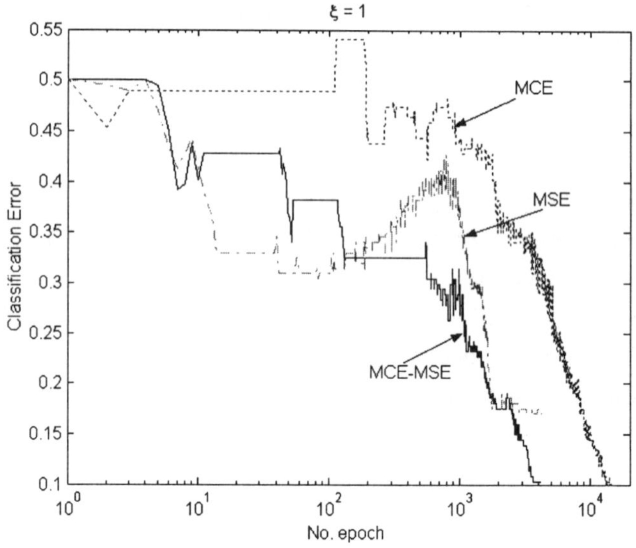

Figure 6.4. A typical convergence curves of different cost functions on the two-spiral classification problem

6.3.3. Implementing MCE-MSE

RBF network is used to illustrate the implementation of MCE-MSE hybrid cost function because of its excellent classification ability compared with feedforward networks. The centers of the RBF networks are firstly initialized. As all data are normalized to the interval [0, 1], the widths of hidden neurons in RBF network are determined by $\sigma = r_0 \times Id$ where Id is the identity matrix, and $r_0 = 0.7$ is used. The linear weights ($W^{(2)}$, $W^{(l)}, B$) of RBF network are randomly initialised with the values between 0 and 1. After initialisation, we can use the learning algorithm listed in the following flow diagram.

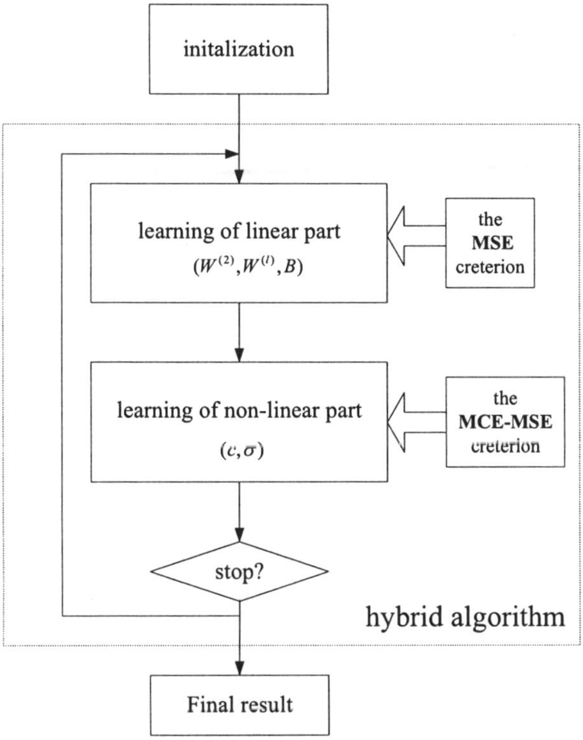

The adjustment of parameters of hidden units is a non-linear process based on the MCE-MSE function Eq. (6.17), whereas the identification of weights between the hidden and the output layers is a linear one based on MSE function Eq. (6.11). Using a simple RBF network, the function can be expressed as

$$Y = f(X) = Z \times W^{(2)} + X \times W^{(l)} + B$$

where Z is the output matrix of the hidden layer. The linear parameters in $f(X)$, i.e., $W^{(2)}$ and $W^{(l)}$, can be calculated by minimizing the MSE function,

$$\begin{bmatrix} W^{(2)} \\ B \end{bmatrix} = \begin{bmatrix} Z \\ 1 \end{bmatrix}^{+} T$$
$$W^{(l)} = X^{+}T \tag{6.18}$$

where $[\cdot]^+$ is defined as the pseudoinverse of $[\cdot]$. Both the MSE function and the MCE function are smooth and differentiable functions, and can be conveniently minimized using gradient descent method. The updating rule of the hidden layer parameter using MCE-MSE function Eq. (6.17) is

$$\Lambda^{(\tau+1)} = \Lambda^{(\tau)} - \frac{1}{2}\left(\Delta\Lambda^{mse} + \Delta\Lambda^{mce}\right)\Big|_{\Lambda=\Lambda^{(\tau)}} \tag{6.19}$$

where Λ^τ denotes the parameter c or σ at the $\tau\, th$ step. The updating rules of Λ based on MSE, i.e., $\Delta\Lambda^{mse}$, have been explained in [3, 8]

$$\left(\Delta c_{jk}^{(\tau+1)}\right)_{mse} = -\eta(\tau)\sum_{i=1}^{N} g_j(x_i)\frac{x_{ik} - c_{jk}^{(\tau)}}{\left(\sigma_{jk}^{(\tau)}\right)^2}\sum_{r=1}^{J} w_{jr}^{(\tau)}(t_{jr} - y_{ir}(\tau)) \tag{6.20}$$

$$\left(\Delta\sigma_{jk}^{(\tau+1)}\right)_{mse} = -\eta(\tau)\sum_{i=1}^{N} g_j(x_i)\frac{(x_{ik} - c_{jk}^{(\tau)})^2}{\left(\sigma_{jk}^{(\tau)}\right)^3}\sum_{r=1}^{J} w_{jr}^{(\tau)}(t_{ir} - y_{ir}(\tau)) \tag{6.21}$$

Simply based on the definition of the MCE function, $\Delta\Lambda^{mce}$ is solved as follows.

$$\left(\Delta c_{jk}^{(\tau+1)}\right)_{mce} = \eta(\tau)\sum_{i=1}^{N} g_j(x_i)\frac{x_{ik} - c_{jk}^{(\tau)}}{\left(\sigma_{jk}^{(\tau)}\right)^2}\sum_{r=1}^{J} w_{jr}^{(\tau)}\frac{\partial E_{mce}}{\partial y_{ir}} \tag{6.22}$$

Setting that

$$\ell_i \equiv \ell(x_i,\Lambda), d_i^{(m)} \equiv d^{(m)}(x_i,\Lambda)$$

according to Eqs. (6.13) and (6.15), we have

$$\frac{\partial E_{mce}}{\partial y_{ir}} = \frac{\partial \ell_i}{\partial y_{ir}} \tag{6.23}$$

$$\frac{\partial \ell_i}{\partial y_{ir}} = \xi\ell_i(1-\ell_i)\frac{\partial d_i^{(m)}}{\partial y_{ir}} \tag{6.24}$$

$$\frac{\partial d_i}{\partial y_i} = -t_i + \left[\frac{1}{M-1}\right]^{\frac{1}{p}}\left[(1-t_i)\times(y_i^p)'\right]^{\frac{1}{p}-1}\times\left[(1-t_i)\cdot\times y_i^{p-1}\right] \tag{6.25}$$

In Eq. (6.25), $\cdot\times$ is a type of matrix operator, and $A\cdot\times B$ is the entry-by-entry product of matrix A and B. The modification rule of the center c based on the MCE-MSE Eq. (6.17) is

$$c_{jk}^{(\tau+1)} = c_{jk}^{(\tau)} - \frac{1}{2}\left(\left(\Delta c_{jk}^{(\tau+1)}\right)_{mse} + \left(\Delta c_{jk}^{(\tau+1)}\right)_{mce}\right)$$

where $\left(\Delta c_{jk}^{(\tau+1)}\right)_{mse}$ and $\left(\Delta c_{jk}^{(\tau+1)}\right)_{mce}$ are given in Eq. (6.20) and Eqs. (6.22-6.25) respectively. In the similar way, the updating rule for the width σ based on MCE-MSE function is

$$\sigma_{jk}^{(\tau+1)} = \sigma_{jk}^{(\tau)} - \frac{1}{2}\left(\left(\Delta \sigma_{jk}^{(\tau+1)}\right)_{mse} + \left(\Delta \sigma_{jk}^{(\tau+1)}\right)_{mce}\right)$$

where $\left(\Delta \sigma_{jk}^{(\tau+1)}\right)_{mse}$ has been given in Eq. (6.21). And $\left(\Delta \sigma_{jk}^{(\tau+1)}\right)_{mce}$ can be delivered as

$$\left(\Delta \sigma_{jk}^{(\tau+1)}\right)^{mce} = \eta(\tau)\sum_{i=1}^{N} g_j(x_i)\frac{(x_{ik} - c_{jr}^{(\tau)})^2}{\left(\sigma_{jr}^{(\tau)}\right)^3}\sum_{r=1}^{J} w_{jr}^{(\tau)}\frac{\partial E_{mce}}{\partial y_{ir}} \quad (6.26)$$

where $\partial E_{mce}/\partial y_{ir}$ is calculated using Eqs. (6.23-6.25).

More thorough tests on the MCE-MSE cost function were reported in other published literature. Three main observations are noticed. First, from the perspective of the classification performance, the MCE-MSE cost function can deliver better classification result than the MSE based method. Second, the MCE-MSE cost function exhibits faster convergence rate than that of the MCE cost function. Even, in certain examples, the efficiency of MCE-MSE based method is substantially larger than that of MCE based cost function. Third, the MCE-MSE function itself exhibits a natural property of faster convergence rate compared with MCE cost function. It is worth noting that the enhanced convergence rate does not require the use of complicated algorithm for adaptively adjusting ξ, which makes it computationally efficient and easy for implementation.

6.4. Feature Selection

Feature selection is an important pre-processing prior classification. It identifies the salient features from the given feature sets for sequential data analysis. In most practical classification and recognition problems, it is very common that a large number of features are given while only a fraction of them are useful to the job. For example, in the problem of sonar signal recognition, data are collected using an array of sensors. But there is not need to use all the collected data from all sensors to classify the sonar signal. Feature selection is to preprocess a given dataset so that we can obtain a smaller subset of representative features for classification or recognition.

In a case of face recognition, it is perceived as a common sense that increasing the number of extracted features may lead to an increase of recognition rate. This, however, does not suggest that we may ever increase the size of feature set, for instance, by extracting irrelevant features. In practice, we experience that the overall performance of recognition is degraded when the feature set is increased beyond certain level. This is due to the problem of *curse of dimensionality*, which was firstly introduced by Bellman at 1961. The issue refers to the exponential growth of hyper-volume as a function of dimensionality.

In neural computing, the problem can be viewed from two perspectives. First, a neural network maps an input space to an output space. A neural network thus requires representing every part of its input space reasonably well in order to know how that part of the space should be mapped. Covering the input space undoubtedly requires resources. It is noted that the amount of required resources is proportional to the hyper-volume of the input space. Including too many irrelevant dimensions causes a network using lots of irrelevant inputs which results performance degradation. When the dimension of an input space is high, a network uses virtually all its resources to represent irrelevant portions of the space. Second, if we include more irrelevant attributes, we increase the size of the input space. This increases the 'distance' between two similar instances. In the case of neural computing, a neural network cannot tell which attributes are irrelevant or not. The network will try to include all of them in the best possible way. This may not be a problem in the learning phase but the network may experience significant problem when it is required to generalize new unseen instances. The network suffers from the problem of overfitting, which means it tries too hard to represent all of the data. As a result, incorrect and highly noisy data presented to it are also included during the network training.

We use the cDNA microarrays Colon cancer classification as an example to illustrate the problem of "*curse of dimensionality*". In the following table, it shows the classification accuracy decreases when more features are presented to the two classifiers.

Table 6.4 Mean Classifying accuracy Colon Tumor data of ten fold tests

No of selected features	KNN	MLP	SVM	Decision tree
50	0.693	0.7234	0.7412	0.6962
500	0.3774	0.6887	0.8226	0.7032
1000	0.3516	0.6758	0.7806	0.7452
2000	0.3484	Not available	0.7516	0.6613

Rather than representing the entire given input features to the output space, feature selection is an important approach to remove the unwanted attributes. When we are selecting the best possible feature subset for performing classification, establishing reliable criteria for evaluating the 'goodness' of feature subsets is crucial. Based on the type of evaluation criteria, feature selection models fall into two main categories – the filter model and the wrapper model. In a wrapper model, the correct recognition rate or error rate of classifiers is directly used for feature selection. Using this type of criteria guarantees high accuracy, but they are usually very computationally demanding. In filter models, various statistics based criteria were developed to measure feature relevance, i.e., the relationship between features and class labels, and/or feature dependence, i.e., the relationship among features. The principle is that the features carrying little or no additional information beyond the selected features are considered redundant and are discarded. There are a number of ways for measuring the feature relevance and dependence. These include linear dependence, correlation coefficients, consistency, and the mutual information (MI) between the selected input variables and output class labels, called MIIO. Before we can further discuss MIIO method, background on the information theory must be included.

6.4.1. Information Theory

Claude Shannon introduced the subject of information theory in 1940's during his work on communication engineering. The development of Information Theory is arguably to be one of the greatest intellectual achievements of the twentieth century engineering work. Information theory has had a pivotal contribution to the 21st century digital communication technology. It has also had a significant impact on applied mathematics, particularly on probability theory. Information theory, originally developed as "A mathematical theory of communication", can provide us with how uncertain a random variable is and how relevant different variables are.

6.4.1.1. Mutual Information

Mutual Information is used for determining the significance of each feature among a given huge feature set for feature selection. But before we move to the concept of mutual information, we need to introduce the concept of entropy. Entropy, originated from thermodynamics, has been extended to statistical mechanics and information theory [2]. For a discrete distribution modeled by $X = \{x_1, x_2, \ldots\ldots, x_N\}$, entropy measures the "information" conveyed by X. The "information" means the uncertainty or the degree of surprise for a particular value of X being drawn. Suppose that x is a value drawn from X, and the event $x = x_k$ occurs with probability p_k, the sum of the probabilities for $x = x_k$ ($k = 1, 2, \ldots\ldots, N$) is 1, i.e., $\sum_{k=1}^{N} p_k = 1$. In the case of $p_k = 1$, there is no uncertainty or surprise for $x = x_k$. A lower value of p_k increases the uncertainty or the "information" when it is known that $x = x_k$ occur. Thus, this "information" is generally measured by $I(x_k) = -log(p_k)$. The "information" contained by the whole event set X is called entropy enumerated by the expected value of $-log(p_k)$, i.e.,

$$H(X) = E(I(x_k)) = -\sum_{i=1}^{N} p_i \log p_i \tag{6.27}$$

A large value of entropy $H(X)$ indicates a high uncertainty about X. When all the probabilities, i.e., p_k for all k, are equal to each other, we have the maximal uncertainty which the value in X is taken, and the entropy $H(X)$ achieves its maximum $\log(1/N)$. Conversely, when all the p_i except one are 0, there is no uncertainty about X, i.e., $H(X)=0$.

When X is a continuous variable, Eq. (6.27) will be extended to

$$H(X) = -\int_{x} p(x)\log(x)dx \tag{6.28}$$

Mutual information was firstly used by Claude Shannon to describe the dependence between two variables. The mutual information between them is zero when the two variables are independent. On the other hand, the mutual information between them is large when the two variables are strongly dependent. There are, however, other interpretations of the mutual information. It can be used to describe the stored information in one variable about another variable. As a result, we will be able to tell the degree of predictability of the second variable when the first variable is known. These interpretations may appear slightly different from the original on, but they are all related to the same issue of dependence and correlation. Mutual information is developed to assess the relation between two variables. Given two variables X and Y, the conditional entropy $H(Y|X)$ measures the uncertainty about C when X is known. And the mutual information (MI) $I(X;Y)$ measures the certainty about Y that is resolved by X. Apparently, the relation of the entropy $H(Y)$, $H(Y|X)$ and $I(X;Y)$ is $H(C) = H(C \mid X) + I(X;C)$,

or, equivalently,

$$I(X;C) = H(C) - H(C \mid X) \tag{6.29}$$

Also, it can be noted that $I(X;Y) = I(Y;X)$. To evaluate the conditional entropy $H(Y|X)$ is

$$H(Y \mid X) = -\int_{x}\int_{y} p(x)p(y \mid x)\log p(y \mid x)dydx \tag{6.30}$$

The MI between X and Y is defined by Shannon as

$$I(X;Y) = \int_y \int_x p(y,x) \log \frac{p(y,x)}{p(y)p(x)} dxdy \tag{6.31}$$

Besides, MI can be seen as the difference between the joint density p(y, x) and the distance p(x)(y). The Euclidean MI is defined as

$$I_e(X;Y) = \int_y \int_x (p(x,y) - p(x)p(y))^2 dxdy \tag{6.32}$$

And based on the Cauchy-Schwartz inequality, i.e., $\log \frac{\|x\|^2 \|y\|^2}{(x^T y)^2} \geq 0$, Principe *et al* extended the Euclidean MI to

$$I_{CS}(f,g) = \log \frac{\left(\int_{x,y} f(x,y)^2 dxdy \right) \left(\int_{x,y} g(x,y)^2 dxdy \right)}{\left(\int_{x,y} f(x,y) g(x,y) dxdy \right)^2} \tag{6.33}$$

where $f(x, y)$ and $g(x, y)$ are $p(x, y)$ and $p(x)p(y)$, respectively.

6.4.1.2. Probability density function (pdf) estimation

Most engineering or physical science research, implicitly or explicitly, assume that the signal and the observation noise are well described as Gaussian random sequences. In most machine learning processes, the underlying probability density functions (pdf) are required for the evaluation of MI and entropy. There are two popular ways to estimate pdf. One is histogram, and the other is mixture model.

Hitherto, the most commonly used technique for estimating the pdf of a continuous-valued random variable is histogram. The idea behind histogram is simple. For a variable x, a set of events $\{x=x_1, x=x_2, \ldots, x=x_n\}$ is given. Histogram defines the underlying probability of x using the appearance frequency. For example, assume that a series of values of x is $\{0, 0, 1, 0, 2, 3, 2, 0\}$. There are totally 8 values collected from x. Base on these values, histogram determines $p(x = 0) = 4/8 = 0.5$, $p(x = 1) = 1/8$, $p(x = 2) = 2/8$, and $p(x = 3) = 1/8$. Generally, histogram is shown as a bar graph. In Fig. 6.5, the histogram of the example is illustrated.

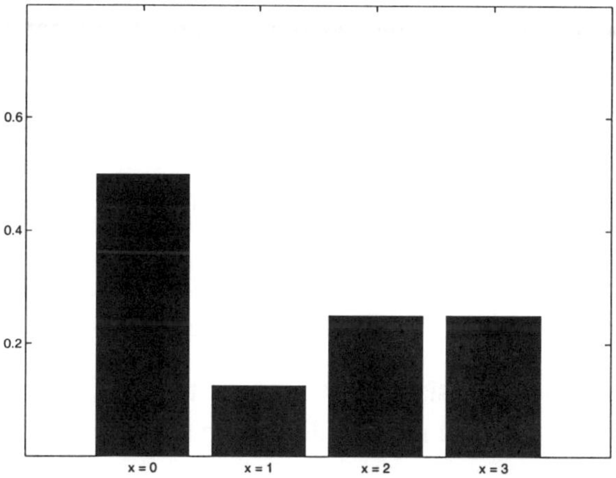

Figure 6.5.

Obviously, histogram can only deal with discrete/categorical variables. For a continuous variable, a discretization process is required. In a 2-dimensional data space, a histogram can be constructed feasibly, but two major problems may be experienced in a high-dimensional data space. First, the increase of data space dimension might have significantly degraded the estimation accuracy due to the sparse distribution of data, especially when the size of data set is relatively small. Second, the required memories exponentially increase with the number of dimensions. This problem may become horrendous when one handle a dataset with huge number of attributes, for instance, an cDNA micrroarrays dataset with over 10 thousand features.

It is widely believed that continuous pdf estimators are more accurate than histograms. Parzen window is a popular mixture probability estimation model, in which all the known data points are used as kernel centres. Given a dataset $X = \{x_1, x_2, \ldots\ldots, x_n\}$, Parzen window estimator is modeled as

$$p(x)=\frac{1}{n}\sum_{i=1}^{n}p(\boldsymbol{x}\mid\boldsymbol{x}_i)=\frac{1}{n}\sum_{i=1}^{n}\kappa(\boldsymbol{x}-\boldsymbol{x}_i,h) \qquad (6.34)$$

where κ is the kernel function of Parzen window, and h is the parameter to determine the width of the window. With the proper selection of $\kappa(\cdot)$ and h, a Parzen window estimator can converge to the real probability density. The kernel function is required to be a finite-value nonnegative function satisfying

$$\int_z \kappa(z)dz=1 \quad \text{and} \quad \lim_{z\to\infty}|z\kappa(z)|=0 \qquad (6.35)$$

The width parameter h is required to decrease until 0 with the increase of the number of the patterns used, i.e., $\lim_{Nq\to\infty} h(Nq)=0$.

A symmetric Gaussian function is a typical choose of κ. The general Gaussian function in an M-dimension space is modelled as

$$\kappa(\mathrm{x}-\mathrm{x}_i,h)=G(\mathrm{x}-\mathrm{x}_i,h)=\frac{1}{(2\pi h^2)^{M/2}|\Sigma_i|^{1/2}}\exp\left(-\frac{1}{2h^2}(\mathrm{x}-\mathrm{x}_i)\Sigma_i^{-1}(\mathrm{x}-\mathrm{x}_i)^T\right) \qquad (6.36)$$

where Σ_i is determined according to the covariance of X. There are many approaches for choosing h. In [7], for instance, it is set that $h=1/\log Nx$, and in [8], $h=(4/(M+2))^{1/(M+4)}n^{-1/(M+4)}$, where n and M are the number of patterns and the dimensionality of dataset X, respectively.

6.4.2. MI Based Forward Feature Selection

Due to the difficulties in high dimensional data spaces, a forward searching is usually adopted. An *MIIO* based forward feature selection algorithm is realised as follows:

1. (Initialization) Set $F\leftarrow$ "initial feature set", $S\leftarrow$ Empty;
2. $\forall f_i \in F$, compute $MIIO(f_i)$;
3. Find the feature f_k that maximizes $MIIO(f_i)$,
   set $F\leftarrow F\setminus\{f_k\}, S\leftarrow\{f_k\}$;
4. Repeat until stopping criterion is met;

5. Calculate $MIIO(S+f_i)$ for all $f \in F$;
6. Choose the feature $f_k \in F$ that maximizes the $MIIO(S+f_i)$;
7. Set $F \leftarrow F \setminus \{f_k\}, S \leftarrow \{f_k\}$;
8. Output the set *S*.

In the above process, the *MIIO* increases gradually. Also, the incremental *MIIO* at each iteration gradually decreases to zero where all the relevant features are selected. Using these properties, the forward process can be stopped when the incremental *MIIO* is small enough because it indicates that the features left in *F* contain very little additional classification information beyond the selected feature set *S*.

Hall (1999) developed a Correlation coefficient based feature selection (CrFS) in which the feature selection criterion, $Corr_s$ is defined as

$$Corr_s = \frac{m\overline{r_{cf}}}{\sqrt{m + m(m-1)\overline{r_{ff}}}} \tag{6.37}$$

where m is the cardinality of *S*, $\overline{r_{cf}}$ is the average feature-class correlation, and $\overline{r_{ff}}$ is the average feature-feature correlation. In the $Corr_s$ Eq. (6.37), the numerator indicates the predictiveness of the feature set *S*; and the denominator measures the extent of redundancy existed in *S*. Apparently, the higher the $Corr_s$ is, the better the feature set for classification will be. For a given feature set, both the irrelevancy features and the redundant features have effect on reducing the value of $Corr_s$. When the $Corr_s$ is determined, the relationship between the individual features and output class labels, and the relationship between two features are considered. Also, the correlation coefficient is a type of linear analysis, in which the nonlinear dependence between two variables may not be measured.

The concept of Consistency based feature selection (CsFS) was introduced to measure the consistency of data described by the feature set *S* with their class labels. $Consistency_s$ is defined as

$$Consistency_s = 1 - \frac{\sum_{i=1}^{J} \text{inconsistency count}}{N} = 1 - \frac{\sum_{i=1}^{J} (n_i^{mj} - n_i^{all})}{N} \quad (6.38)$$

where, for a feature subset *S*, *J* is the number of distinct data patterns described by *S*, n_i^{all} is the number of occurrences of the ith distinct pattern, n_i^{mj} is the largest number of occurrences of the ith distinct pattern in the same class, N is the number of all patterns. To support that, there is a group of *n* patterns in which all patterns match with each other without considering their class labels. In this pattern group, n_1 patterns belong to class 1, n_2 to class 2, n_3 to class 3, where $n=n_1 + n_2 + n_3$. If $n_3 = \max(n_1, n_2, n_3)$, the inconsistency count of this pattern group is $(n - n_3)$. A higher value of $Consistency_s$ indicates that the distribution of data patterns defined by *S* is more similar to that of the class labels.

CsFS is sensitive to noise data. Also, in order to calculate the $Consistency_s$, continuous input variables should be discretized in a way similar to produce histograms. CsFS suffers the same problem of histograms that the pattern shortage substantially degrades the performance of CsFS. It is worth noting that $Consistency_s = 1$ may not be a reliable indication that the best feature subset is selected. In a high-dimensional space, there are hardly sufficient matching patterns due to the sparse distribution of data patterns. From Eq. (6.38), $Consistency_s = 1$ can be obtained because $n_i^{mj} - n_i^{all} = 0$ for any *i*.

6.4.2.1. MIFS and MIFS-U

Battiti (1994) and Kwok and Choi (2002) employed *MIIO* to select feature. As shown in Fig. 6.6, because of the high-dimensional problems experienced by histograms, Battiti and Kwok did not direct estimate *MIIO* ($I(S + f_i; C)$) represented by the area $A_1+A_2+A_3$. Instead, they analyzed $I(f;C \mid S)$ represented by the area A_1 through the 2-dimensional MI. These methods are acceptable because the area A_2+A_3 is common for $f_i \in F$. To analyze $I(f;C \mid S)$ represented by the area A_1, Battiti's MIFS used

$$I(f;C)-\beta\sum_{\dot{f}\in S}I(f;\dot{f}) \tag{6.39}$$

and Eq. (6.40) was used in Kwok's MIFS-U.

$$I(f;C)-\beta\sum_{\dot{f}\in S}I(f;\dot{f})\frac{I(C;\dot{f})}{H(\dot{f})} \tag{6.40}$$

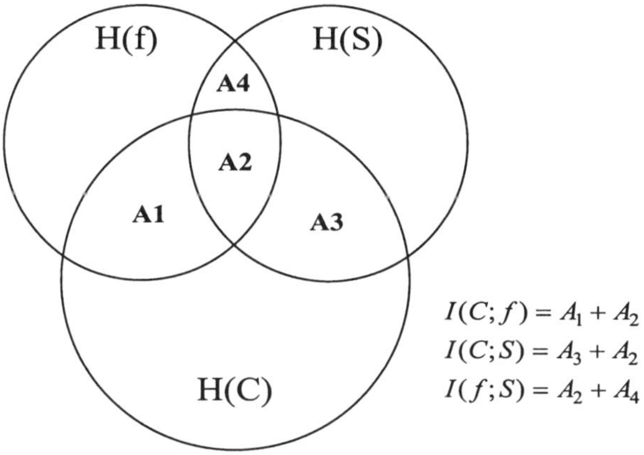

Figure 6.6. The relation between input feature variables and output classes in terms of MI

This indirect analysis on high dimensional MI generates two major problems, i.e., sub-optimization and no estimate on the *ANSF*. Also, these methods cannot provide a principled guide to reduce the redundancies. In Eqs. (6.39) and (6.40), the parameter β is aimed at regulating the relative importance of MI between f and S with respect to the MI of f with output classes. Also, β determines the capability of handling the redundant features. Despite the importance of β, there is no guideline provided for its selection. The selection of β is rather problem dependent and does require strong experience.

6.4.2.2. Using quadratic MI

Given a classification task with class labels *C*, and the selected feature set *S*, the criterion *MIIO* of a feature f_m ($f_m \notin$ S) is defined as,

$$MIIO(f_m) = I(S + f_m; C) \tag{6.41}$$

The *MISF* measures the similarity between the feature subset *S* and a single feature f_m. The similarity between *S* and f_m ($f_m \notin S$) can be estimated by $I(S;f_m)$. As addition of input variables have no effect on decreasing the MI (Cover & Thomas, 1994), $I(S;f_m)$ has to be large when the MI of f_m with any feature in *S* is large. That is, if f_m is very similar to f_i ($f_i \in S$), f_m should be redundant to S. Hence, given a feature set *S*, the *MISF* of f_m ($f_m \notin S$) is defined as

$$MISF(f_m) = \arg\max_{f_i \in S} \left(\frac{I(f_m; f_i)}{H(f_i)} \right) \tag{6.42}$$

When $MISF(f_m;S) \geq \theta$, the feature f_m is considered as a redundant feature to *S*. Hence, the feature f_m should not be added into *S* accordingly. Using the definition of entropy and MI, we have $H(f_i) = H(f_i \mid f_m) + I(f_i; f_m) \geq I(f_i; f_m)$. Hence $\theta \leq 1$ with the equality if and only if f_m contains no additional information beyond f_i ($f_i \in S$). In general, θ is chosen as 0.5.

MIIO and *MISF* are estimated in a similar way of using the quadratic MI and the Gaussian based density estimators. In order to further reduce the computational complexity, a supervised data compression algorithm (Huang and Chow, 2003) is used. This data compression algorithm firstly clusters the whole data points {*X, C*} into homogenous subgroups $U = \{u_1, u_2, ..., u_{Nu}\}$. Following the clustering process, certain data points are sampled for each cluster. Let the data set $SX = \{sx_1, sx_2, ..., sx_{Ns}\}$ be the compression result of {*X, C*}. For the cluster u_j, let n_j^x be the number of data points in the original data {*X, C*} and n_j^s be the number of data points in the compressed data *SX*. Obviously, we have $n_j^x \geq n_j^s$ due to the data compression process.

With the data set *SX*, conventional Parzen window estimator (Parzen, 1969) assumed that

$$p_p(sx_i) = \frac{1}{N_s} \tag{6.43}$$

In order to make use of more information contained in the original data set, different from conventional Parzen window estiamtor, we estimate $p(sx_i)$ using

$$p(sx_i) = \frac{n_j^x}{Nx} \times \frac{1}{n_j^s} \tag{6.44}$$

Hence, we estimate the marginal density and conditional density as

$$p(x) = \sum_{i=1}^{Ns} p(sx_i)p(x \mid sx_i) = \sum_{i=1}^{Ns} p(sx_i)\kappa(x - sx_i, \Sigma_i) \tag{6.45}$$

$$p(x \mid l) = \sum_{sx_i \in class\, l} p(sx_i)p(x \mid sx_i) = \sum_{sx_i \in class\, l} p(sx_i)\kappa(x - sx_i, \Sigma_i) \tag{6.46}$$

$$p(l) = \sum_{sx_i \in class\, l} p(sx_i) \tag{6.47}$$

where $L = \{l_1, l_2,,,,, l_{Nl}\}$ is the class label set. Actually, the proposed density estimator is a generalization of the Parzen window density estimator. In our proposed density estimator, a symmetric Gaussian function is chosen as kernel function. The Gaussian function in *M* dimension is

$$\kappa(z - z_0, \Sigma) = G(z - z_0, \Sigma) = \frac{1}{(2\pi h)^{M/2} \|\Sigma\|^{1/2}} \exp\left(-\frac{(z - z_0)^T \Sigma^{-1} (z - z_0)}{2h^2}\right) \tag{6.48}$$

where Σ is determined from the variance matrix of the overall data, *h* is the bandwidth of kernel function. In general, $\Sigma = I$ is set. There are many methods for choosing *h*. The method developed by Silverman (1986) is used in this paper, i.e., *h* in Eq. (6.48) is calculated by

$$h=\left\{\frac{4}{(M+2)}\right\}^{1/(M+4)} Nx^{-1/(M+4)} \tag{6.49}$$

The quadratic MI (2.6) can be extended to discrete variables as

$$I_{CS}(X;C)=\log\frac{(\sum_c \int p(x,c)^2 dx)(\sum_c p(c)^2)(\int p(x)^2 dx)}{(\sum_c p(c)\int p(c,x)p(x)dx)^2} \tag{6.50}$$

Using the property of Guassian function, the quadratic MI Eq. (6.50), and the marginal pdfs and conditional pdfs Eqs. (6.45-6.47), we are able to estimate *MIIO* with

$$MIIO=I_{CS}(X;C)=\log\frac{V_{(c,x)^2}V_{(c)^2}V_{(x)^2}}{V_{(cx)}{}^2} \tag{6.51}$$

Where,

$$V_{(c,x)^2}=\sum_l\int p(l,x)^2 dx=\sum_{k=1}^{Nl}\sum_{sx_i\in \text{class } l_k}\sum_{sx_j\in \text{class } l_k} p(sx_i)p(sx_j)G(sx_i-sx_j,2I) \tag{6.52}$$

$$V_{(c)^2}=\sum_{k=1}^{Nl}p(l_k)^2=\sum_{k=1}^{Nl}\left(\sum_{sx_i\in \text{class } l_k}p(sx_i)\right)^2 \tag{6.53}$$

$$V_{(x)^2}=\int p(x)^2 dx=\sum_{i=1}^{Ns}\sum_{j=1}^{Ns}p(sx_i)p(sx_j)G(sx_i-sx_j,2I) \tag{6.54}$$

$$\begin{aligned}V_{(cx)}&=\sum_l\int p(l,x)p(l)p(x)dx\\&=\sum_{k=1}^{Nl}\left(\sum_{sx_i\in \text{class } l_k}p(sx_i)\right)\sum_{j=1}^{Ns}\sum_{sx_i\in \text{class } l_k}p(sx_j)p(sx_i)G(sx_j-sx_i,2I)\end{aligned} \tag{6.55}$$

Only MIs between two continuous variables are required when estimating *MISF* because $H(f_i)=I(f_i;f_i)$. For the input data containing two input features f_a and f_b, assume that the result obtained by the data compression algorithm is

$$SX = \{(sx_{a1}, sx_{b1}), (sx_{a2}, sx_{b2}), \ldots\ldots, (sx_{aN_s}, sx_{bN_s})\}$$

the 2-dimensional MI $I(f_a; f_b)$ can be estimated with

$$I(f_a; f_b) = \log \frac{V_{(f_a,f_b)^2} V_{(f_a)^2} V_{(f_b)^2}}{V_{(f_{ab})}^{\ 2}} \tag{6.56}$$

where

$$V_{(f_a,f_b)^2} = \sum_{i=1}^{Ns} \sum_{j=1}^{Ns} \prod_{q=1}^{2} p(sx_i)p(sx_j)G(sx_{qi} - sx_{qj}, 2I) \tag{6.57}$$

$$V_{(f_q)^2} = \sum_{i=1}^{Ns} p(sx_i) \sum_{j=1}^{Ns} p(sx_j)G(sx_{qi} - sx_{qj}, 2I) \tag{6.58}$$

$$V_{(f_{ab})} = \sum_{i=1}^{Ns} p(sx_i) \prod_{q=1}^{2} \left(\sum_{j=1}^{Ns} p(sx_j)G(sx_{qi} - sx_{qj}, 2I) \right) \quad q = a, b \tag{6.59}$$

The forward feature selection is an iterative process. It begins with an empty feature set and additional features are included one by one. The forward feature selection process can be implemented as follows.

Step 1. Set F $\leftarrow$ "initial feature set", S $\leftarrow$ Empty, the number of selected features $j = 0$;

Step 2. If cardinality (F) > 0, then compute $MIIO(f_i)$ for $\forall f_i \in F$;
Otherwise goto Step 6;
End If;

Step 3. Choose the feature $f_k \in F$ that
$MIIO(f_k) = \arg\max_{f_i}(MIIO(f_i))$;

Step 4. Identify the redundancy:
If $MISF(f_k) \geq \theta$, then $F \leftarrow F \backslash \{f_k\}$, goto Step 3;
Otherwise $F \leftarrow F \backslash \{f_k\}, S \leftarrow \{f_k\}$, $j = j + 1$,
$MIIO_j = MIIO(f_k)$, goto Step 5;
End If;

Step 5. Stop the iterative process:
If $(MIIO_j - MIIO_{j-1}) / MIIO_1 \leq \gamma$, goto Step 6;
Otherwise goto Step 2;
End If;

Step 6. Estimate the appropriate number of the selected features (ANSF). $B \leq ANSF \leq j$, where B satisfies
$MIIO_B \geq \lambda \times \arg\max_i(MIIO_i)$;

Step 7. Output the set S and ANSF.

In the above stopping criteria, γ should be as close to zero as possible in order to assure most of the original classification information are preserved in the selected features. Generally, $\gamma = 0.01$ is considered a small enough index for delivering satisfactory results. The parameter λ determines the range of estimates. It is well known that the optimal feature set is rather classifier dependent. At last, we can say that one can use the above feature selection scheme for effectively eliminating huge redundancy features in a huge feature set. This is useful in many nowadays physical and engineering applications.

Exercise

Q6.1. The average entropy $H(d)$, is a binary function. It is given that $H(d) = -d\log_2 d - (1-d)\log_2(1-d)$, determine

(a) $H(0.5)$;
(b) The average entropy $H(d)$ when the probability d is uniformly distributed in the range $0 \leq d \leq 1$.

Q6.2. To understand the probability and conditional probability.
Based on the following knowledge, calculate the probability that if somebody is tall, that person must be male. It is known that the probability of being male is P(M) = 0.5, and the probability of being female is P(F) = 0.5. Also, it is known that 20% of males

are tall, that is p(T|M) = 0.2, and p(T|F) = 0.06. This question asks you to calculate p(M|T).

Q6.3. To understand the entropy, conditional entropy and mutual information.
To win a volleyball game, a team requires get 3 out of 5 sets. Thus, a game comes to end once either side has won 3 sets. Assume that we have two teams, A and B. The final result may be AAA, ABBB, BAABB, etc. Let X be the final competition result, and Y be the number of sets played. Also, it is known that A and B are equally matched and the results of the sets are independent. Calculate H(X), H(Y), H(X|Y), H(Y|X) and I(X;Y).

Q6.4. Coin flip problem. Flip a coin until the first head occur. Let X be the number of flips conducted. Calculate the probability of X = i. Then based on the probabilities, find H(X) in bits. The following equations will be useful

$$\sum_{n=1}^{\infty} r^n = \frac{r}{r-1} \text{ and } \sum_{n=1}^{\infty} nr^n = \frac{r}{(r-1)^2}$$

Q6.5. Assume that X and Y are random values respectively from {x1,x2, …, xn} and {y1, y2, …, ym}.

a) Let Z = aX where a is a constant. Show that H(X) = H(Z);
b) Let Z = X + Y. Show that H(X|Y) = H(Z|Y);
c) Let Z = X + Y. Show that if X and Y are independent, H(Z) ≥ H(X) and H(Z) ≥ H(Y);
d) Let Z = X + Y. Given a condition (i.e., the relationship between X and Y) in which H(Z) < H(X) and H(Z) <H(Y);
e) Let Z = X + Y. Under what condition does H(Z) = H(X) + H(Y).

Chapter 7

Engineering Applications

We have provided the general framework of Neural Computing and its related theories in the previous six chapters. In this Chapter, we aim to demonstrate the applications of the three selected areas. The first one is the use of feedforward neural network on a widely interested area of time-series based forecasting problem. Many literatures on this topic are published. In this Chapter, we introduce the way of including weather parameters, which are considered as noises to the model, and model selection to short term electric load forecasting problem. The second application example is the use of Self-Organizing Map on an image retrieval system. This is a useful and interesting application to the IT area. The last example is about the use of feature selection method for bioinformatics area. Bioinformatics has become a very important research area which requires a lot of support from neural computing. Computation intelligence methods have once again shown its versatility and importance on gene selection and cancer diagnosis. This example will surely provide new insight on neural computing.

7.1. Electric Load Forecasting

System load forecasting is an essential function in power system control centers. Short-term load forecasting (STLF) is an indispensable procedure in the real-time control of power generation and efficient energy management systems. It is used for establishing the power station operation plan and the unit operation plan, together with generation and spinning reserve planning of energy exchange. Significant forecast errors can result in either excessively conservative scheduling or excessively risky scheduling that can cause heavy economic penalties. Large savings of money and energy can be achieved if accurate load forecasts are used to support these planning and scheduling.

Interest in applying neural network to electric load forecasting has begun since 1990. Feedforward neural network is used to incorporate the previous load demands, day of week, hour of day and temperature information for load forecasting. There are other extended work including additional input variables of a seasonal factor and cooling/heating degree is included in a single neural network. Most neural models treat electric load demands as a nonstationary time-series and they modeled the load profile by a recurrent neural network.

Amongst the above neural based forecasting techniques, most of them can be classified as nonlinear time-series approaches assuming that the load can be decomposed into two components, namely weather dependent factor and weather independent factor. The behavior of weather independent load is mostly characterized by seasonal factors and trend profiles in terms of time. The weather dependent load is often estimated through weather variables such as the temperature, relative humidity and wind speed.

Up to now, most neural-based techniques consider electric load demands as "short-time" wide sense stationary. In general, load profile, however, behaves as a nonstationary time-series, especially in developing countries. This reason makes those techniques difficult to provide accurate forecasts over time. Besides, those load forecast models can be summarized in the following two equations. For static model (feedforward neural network),

$$x_{t+l} = h(x_{t-1}, x_{t-2}, \ldots, x_{x-p}, \hat{w}_{t+l}, w_{t-1}, w_{t-2}, \ldots, w_{t-q}) + e_{t+l} \quad (7.1)$$

and, for dynamical model (recurrent neural network),

$$x_{t+l} = h(x_{t-1}, x_{t-2}, \ldots, x_{x-p}, \hat{w}_{t+l}, w_{t-1}, \ldots, w_{t-q}, e_{t-1}, \ldots, e_{t-r}) + e_{t+l} \quad (7.2)$$

where xt is the load consumption, $\hat{w}_{t+l}$ is the weather forecast, wt is the weather information and et is noise residual at time t. The nonlinear function h is nonlinearly approximated by a neural network. From the above models, the load forecast $\hat{x}_{t+l}$ is mainly estimated based on the current available load profile. We can rewrite Eq. (7.1) in the form of

$$\begin{aligned} x_{t+l} &= x_{t+l-24} + \Delta x_{t+l} \\ &= h(x_{t-1}, x_{t-2}, \ldots, x_{x-p}, \hat{w}_{t+l}, w_{t-1}, w_{t-2}, \ldots, w_{t-q}) + e_{t+l} \end{aligned} \quad (7.3)$$

Because the load xt+l and xt+l-24 are highly correlated, the deviation xt+l is commonly about 2-3% of the base load for weekdays. Actually, our target of load forecasting is aimed at the prediction of the deviation xt+l instead of the load xt+l. In other words, only the 2-3% dynamic range of a neural network is used for prediction in those models. It is difficult to have an overall forecast error less than 2% when the data is perturbed by weather disturbance that strongly influences the behavior of power consumption, especially in higher population areas, such as Hong Kong, and the dynamic range of the neural network is not fully utilized.

7.1.2. Nonlinear Autoregressive Integrated Neural Network Model

Now, we focus on a multilayer feedforward neural network (FNN) and how it may be used to forecast hourly load consumption of the coming day. It has been proved that an FNN can be used as a universal function approximator (Hornik et al. 1989). A function defined on a compact set in C[a,b] or Lp[a,b] can be approximated arbitrarily well by an FNN with one hidden layer (Chen 1993). However, FNN does not always provide acceptable performance in applications of time-series prediction. In many time-series predictions, the time-series model is always based on nonlinear autoregressive (NAR) models which is

$$x_t = h(x_{t-1}, x_{t-2}, \ldots, x_{t-p}) + e_t \quad (7.4)$$

The neural STLF models (Peng et al. 1993, Lu et al. 1993, Chen et al. 1992, Djukanovic, et al. 1993, Paplexopoulos et al. 1994) can be considered as a modified NAR model Eq. (7.1) which is given by

$$x_{t+l} = h(x_{t-1}, x_{t-2}, \ldots, x_{x-24}, w_t) + e_{t+l} \quad (7.5)$$

and the unknown smooth function h is nonlinearly approximated by FNN. Hence, the neural optimal predictor is given by

$$\hat{x}_{t+l} = \sigma\left(\sum_{i=0}^{H} W_i^1 \sigma\left(\sum_{j=0}^{24} W_{ij}^0 x_{t-j} + \sum_{k=1}^{m} W_{ik}^0 w_k^t\right)\right) \tag{7.6}$$

where $0 \le l \le 24$ and the function is a smooth function bounded monotonic function, tanh(0.5x). The vector wt of m components contains the available weather information at time t. The parameters W_{ij}^0, W_{ik}^0 and W_i^1 are the neural network weights.

For stationary time-series, the upper and lower bounds of value xt can be estimated because the statistical characteristics of stationary time-series are time-invariant. Based on the results of (Chen et al. 1993), the unknown function h can be approximated arbitrarily well by FNN. In contrast, the statistical parameters of nonstationary time-series, such as E{xt}, are time-variant. The upper and lower bounds of value xt can hardly be found so that the NAR model is not the most appropriate model for neural time-series prediction of nonstationary time-series. To obtain accurate load forecasting, the most appropriate model must be identified in accordance with the nature of load consumption. The electric load consumption is actually nonstationary. A modified Nonlinear AR integrated (NARI) model is used for STLF. Several important weather factors are also included in the model because weather variation is one of the crucial disturbances to electric load demand. Consequently, the modified NARI model for STLF is given by

$$\hat{x}_{t+l} = x_{t+l-d} + \hat{h}(x_{t-1}, \ldots, x_{t-24}, w_t) + e_{t+l} \tag{7.7}$$

and the neural optimal predictor is then formulated by

$$\hat{x}_{t+l} = x_{t+l-d} + \sigma\left(\sum_{i=0}^{H} W_i^1 \sigma\left(\sum_{j=0}^{24} W_{ij}^0 x_{t-j} + \sum_{k=1}^{m} W_{ik}^0 w_k^t\right)\right) \tag{7.8}$$

where $0 \le l \le 24$. The weather information vector w_t contains the following components:

a. Temperature:

Current day: 24 hourly temperatures, maximum temperature, minimum temperature, mean temperature.

Next day (forecast): maximum temperature, minimum temperature.

Temperature is the most important weather variable. System load is rather sensitive to temperature changes. Large system load change occurs at the time with large temperature rise and fall.

b. Relative Humidity:

Current day: 24 hour relative humidity

In summer, for a given range of temperature, relative humidity is significant in affecting the utilization of air conditioning. In addition, sky cover and probability of rain can be related to the relative humidity.

c. Rainfall:

Current day: rainfall

Rainfall is the other crucial parameter affecting system load. It has never been addressed in other works. In Hong Kong, rain is a common phenomenon, especially in summer. It directly affects air conditioning load and lighting load because relative humidity and sky cover are all related to rainfall.

d. Sunshine:

Current day: sunshine

Sunshine is an index measuring the degree of sky cover that directly affects lighting load consumption.

The architecture of the neural network model is illustrated in Fig. 7.1(b). The structure of the modified NAR model for STLF is also depicted in Fig. 7.1(a) Compared to the weather dependent component of electric load, the weather independent component, in fact, exhibits low frequency characteristics. According to Eq. (7.8), the low frequency weather independent component can be excluded because the operator $\nabla^d x_t$ behaves as a high-pass filter. Hence, the NARI neural network model is called a weather compensation neural network because the weather dependent component will only be determined using weather information and load consumption of the previous day.

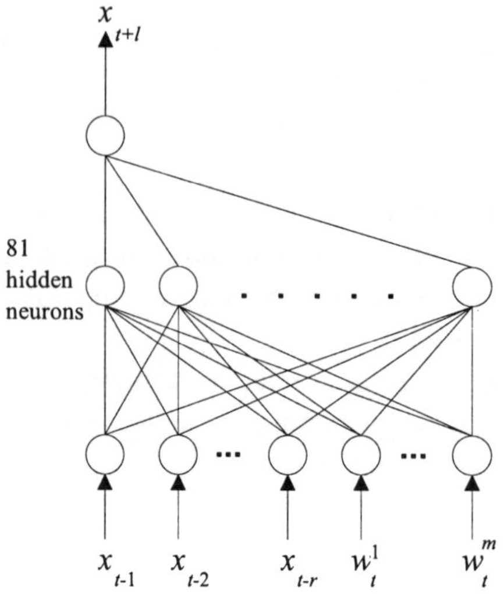

Figure 7.1. (a) The architecture of modified Nonlinear AR (NAR) model for neural short-term load forecasting

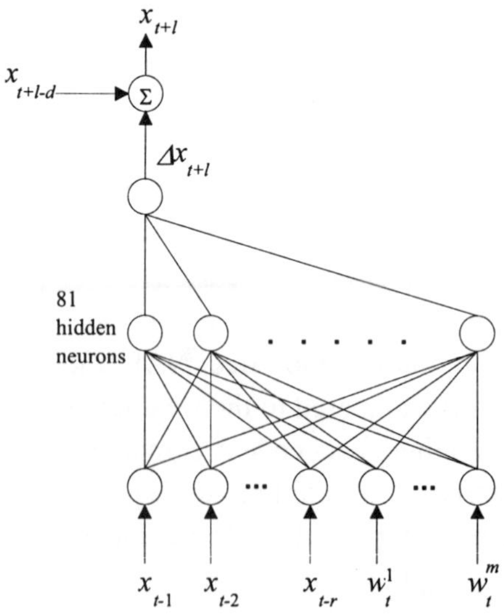

Figure 7.1. (b) The architecture of weather compensation neural network (modified NARI model)

7.1.3. Case Studies

Case studies were carried out for a one-day ahead forecasting of hourly electric loads of weekdays using historical data on Hong Kong Island. The results were analyzed based on the following indices:

Standard Deviation:

$$\sigma = \sqrt{\frac{1}{N}\sum_{i=1}^{N}\left(x_t^i - \hat{x}_t^i\right)^2} \tag{7.9}$$

Percentage Error:

$$\varepsilon = \frac{1}{N}\sum_{i=1}^{N}\frac{\left|x_t^i - \hat{x}_t^i\right|}{x_t^i}100 \tag{7.10}$$

Two different neural network models of NAR, Eq. (7.6) and NARI, Eq. (7.8) are established using FNN with a single hidden layer. For simplicity of terminology, NAR stands for neural network NAR model while NARI represents neural network NARI model. An adaptive backpropagation learning algorithm of extended backpropagation described in Chapter 2 is used because this algorithm can significantly speed up the training speed by adaptively tuning the learning rate and the momentum factor. The same structure of FNN with 81 hidden neurons is applied to these two load models for one-day ahead hourly load forecasting. The 24 hour load consumption of the previous day as well as the weather information is used as input variables.

The influence of weekends and standard holidays on the load is not considered in the STLF. Before the NARI neural network model is built, the parameter d has to be estimated. The selection of parameter d is based on the criterion of minimizing $E\left\{\left|\nabla^d x_t\right|\right\}$. The value of $E\left\{\left|\nabla^d x_t\right|\right\}$ is directly related to the correlation between x_t and x_{t-d} because $E\left\{\left|\nabla^d x_t\right|\right\}$ will be smaller if x_t and x_{t-d} are more correlated. Therefore, d is estimated based on the correlation between x_t and x_{t-d}.

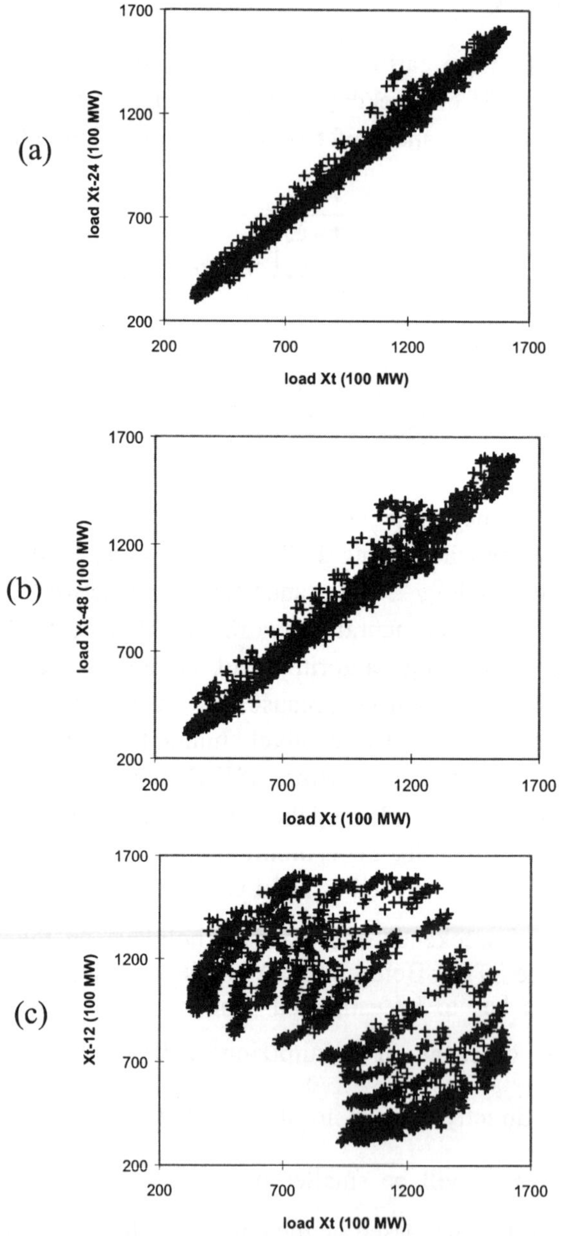

Figure 7.2. The scatter plot (a) between loads x_t and $x_{t\text{-}24}$ (b) between loads x_t and $x_{t\text{-}12}$ (c) between loads x_t and $x_{t\text{-}48}$

The scatter plot of x_t and x_{t-d} enables the correlation between the two variables to be found. As a result, whether x_t and x_{t-d} are linear or nonlinear correlated, the correlation can then be visually determined. In Fig. 7.2(b), the points between x_t and x_{t-12} scatters all over the diagram, which implies x_t and x_{t-12} are not correlated. Apparently, Figs. 7.2(a) and 7.2(c) indicate that they are more correlated. Compared to Figs. 7.2(a) and 7.2(c), x_t and x_{t-d} in Fig. 7.2(a) appears to be the most correlated. Hence, the parameter d is chosen to be 24. In our investigation, the load demands of weekdays from March 12 to August 26, 1992 were used for testing while the load demands of weekdays in 1991 were used for the training set. Figs. 7.3 to 7.4 show the results of model NAR and NARI respectively. Fig. 7.3 illustrates the weather effect on the daily load profile which is defined by the following equation

$$E_w = \frac{1}{24}\sum_{i=1}^{24}\frac{\left|x_t^i - x_{t-1}^i\right|}{x_t^i}100 \tag{7.11}$$

Due to the seasonal transition, the load profile from day 5 to day 20 changed severely as shown in Fig. 7.3. NAR and NARI can predict load consumption with acceptable accuracy during this interval. Outside the seasonal transition, NAR, however, cannot provide more accurate forecast as illustrated in Figs. 7.3 and 7.4. In contrast, Figs. 7.3 and 7.5 manifest that NARI can forecast the electric load at comparable accuracy compared to the case for seasonal transition. Table 7.1 summarizes the overall percentage errors. NARI can provide the most accurate load forecast and the overall percentage error is 1.755 for the forecasts of 24-hours ahead. The overall percentage error of STLF using NARI can be reduced by 0.65%.

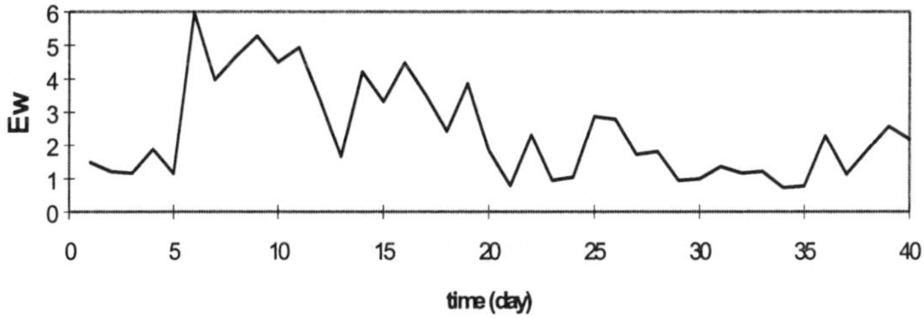

Figure 7.3. The effect of load change due to the weather change

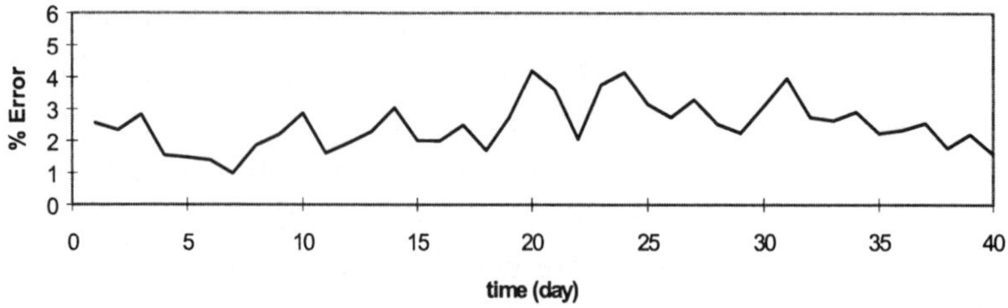

Figure 7.4. Percentage error of load forecast using neural network NAR model

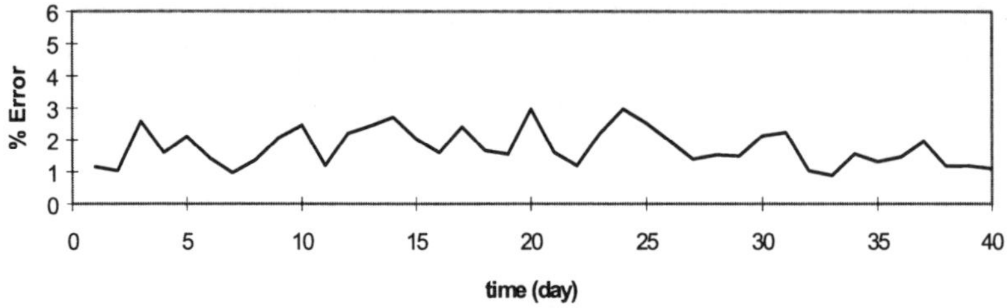

Figure 7.5. Percentage error of load forecast using neural network NARI model

Table 7.1 Comparison of forecasting results for March 12 - August 26, 1992

	NAR	NARI
overall % Error	2.492	1.755

It must be noted that the target of load forecasting is only aimed at the prediction of the small deviation Δx_{t+l} instead of the load x_{t+l} based on Eq. (7.3). In model NAR, only the 2-3% dynamic range of FNN is used for prediction. Apparently, the overall forecast error can hardly be less than 2% when the data is perturbed by the weather disturbance that strongly influences on the behavior of power consumption, especially in Hong Kong. The neural network model can utilize the whole dynamic range of the output for forecasting the target Δx_{t+l}. From Table 7.2, the standard deviation of forecast error of NARI can significantly be reduced in almost all time slots by 1190 MW, which is about 0.7 % of the peak load.

Table 7.2 Comparison of forecasting results of each hour

	NAR		NARI		NARI (136-1)	
hour	% Error	St.Dev. /100 MW	% Error	St.Dev. /100 MW	% Error	St.Dev. /100 MW
1	1.522	11.607	1.274	11.134	1.214	11.479
2	1.870	14.068	1.181	9.303	1.059	8.457
3	2.302	18.393	1.391	11.335	1.280	10.215
4	2.198	15.356	1.229	8.654	1.275	9.512
5	2.268	18.014	1.353	9.545	1.348	9.509
6	2.044	19.032	1.308	9.366	1.354	9.767
7	1.157	14.256	1.747	16.361	1.417	13.051
8	1.138	19.399	1.298	18.138	1.371	17.824
9	2.168	36.059	1.763	27.909	1.314	22.063
10	2.789	49.709	1.719	28.621	1.139	21.040
11	2.916	52.315	1.733	30.518	1.188	22.101
12	3.224	60.484	1.946	34.929	1.271	24.643
13	2.864	55.318	1.549	29.621	1.366	24.637
14	2.805	52.329	1.801	33.888	1.283	25.867
15	2.626	50.027	1.707	33.280	1.483	27.379
16	2.792	52.377	1.949	37.454	1.489	28.359
17	2.853	50.874	1.443	28.310	1.418	25.561
18	2.262	37.006	1.807	31.220	1.685	28.139
19	2.875	43.099	2.075	34.141	1.785	28.673
20	2.819	41.576	2.133	33.299	1.882	31.244
21	2.716	37.533	2.215	32.991	2.014	30.564
22	3.338	41.282	2.312	34.541	1.950	30.979
23	3.296	36.489	2.538	31.960	2.195	28.631
24	2.956	35.043	2.642	29.287	2.324	25.566
overall	2.492	39.051	1.755	27.146	1.504	22.856

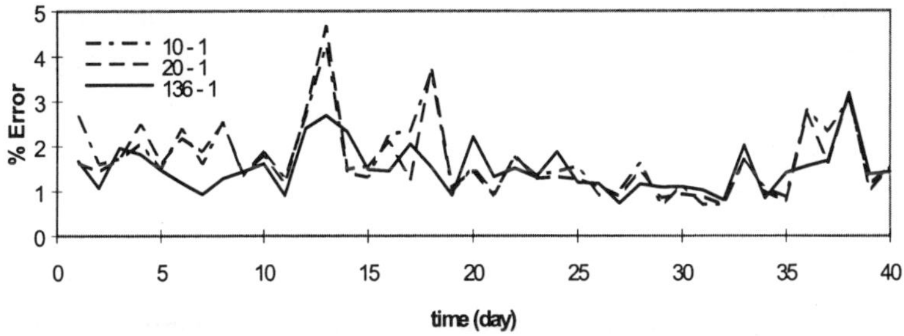

Figure 7.6. Comparison of percentage error for different sizes of moving window using neural network NARI model

The characteristic of electric load consumption gradually changes because of many uncontrollable factors. The weather compensation neural network (NARI model), without keeping track of the change of load characteristic, will degrade the forecasting performance in later years. An adaptive tracking scheme can be employed so that the weather compensation neural network can be retrained every day. This scheme can efficiently update the neural network to adapt the changing conditions of the environment. Further details on this issue can be referred to other literatures.

7.2. Content-based Image Retrieval Using SOM

Content-based image retrieval (CBIR) is one of the effective techniques for retrieving semantically relevant images from unlabelled image data sets based on automatically extracted features. It has been an ongoing research subject for more than a decade (Rui et al. 1999). It usually retrieves relevant images based on the image comparison of visual contents, such as color, texture, shape, structure, etc. A region-based CBIR system using a growing hierarchical self-organizing quadtree map (GHSOQM) is described. Each image in the CBIR system is first segmented into several regions. Each region has similar features for colors and textures. Each image is thus represented by a region-based feature matrix. Different images may have different number of regions. As far as we are aware, there, hitherto, has not been a definition of feature matrices for neurons in neural networks. All neurons in GHSOQM have a fixed number of row vectors in feature matrices, which mean that all neurons represent images with fixed number of regions in feature space. A new criterion for image distance can be applied to region-based representation of images. GHSOQM organizes images in hierarchical levels. Since SOM usually defines an elastic topology-preserving net stretched in the input space, high-dimensional images can be arranged in a 2-D grid at different precision level in GHSOQM. Images belonging to neighboring neurons have similar semantic meanings in an SOM at the same level. Dead or useless neurons at each of the hierarchical level are removed. This is significant because the storing space for neurons is saved and searching time can be significantly speeded up. Coupled with relevance feedback technique, the CBIR system can achieve a good retrieving result.

7.2.1. GHSOQM Based CBIR Systems

7.2.1.1. Overall Architecture of GHSOQM-Based CBIR System

All images are processed by the same feature extraction method. Each image is first segmented into several similar regions, such as by the JSEG algorithm (Deng et al. 1999). The characteristic features, i.e., colors and textures, are extracted for each region of an image. After all available images are processed; GHSOQM is trained by using region-based feature vectors for images. After completion of training, all images are first assigned to the SOM at the first level according to the nearest distance. Then the images assigned to a neuron at the first level are assigned to the child neurons of the neuron. The assignment process proceeds until the leaf neurons are assigned with images. After completion of image assignment to neurons, the GHSOQM-based CBIR system is ready for query or retrieval. The image retrieval procedure can be described as the following steps:

Step 1) A submitted query image is processed to extract region-based features.
Step 2) The CBIR system first finds a nearest neuron at the top level of GHSOQM.
Step 3) If the number of associated images in the nearest neuron exceeds a prespecified minimum number λ, find a nearest child neuron of the nearest neuron at the next bottom level.
Step 4) Repeat the step 3 until the found neuron is associated with the least number of images that is still more than the prespecified number λ. The last found neuron is a target one for next steps.
Step 5) Directly compare the distance between the query image and the target neuron by region-based features. Sort the images by distance with an ascending order and provide them to users.
Step 6) Users select some retrieved images as relevant ones. This information is feedback to front-end of the CBIR system. The

old query is then modified to a new one according to the users' feedback. And the new query is supposed to retrieve more relevant images. This step is called relevance feedback (RF). RF is usually iterated for several times.

The architecture of GHSOQM-based CBIR system is shown in Fig. 7.7. Note that the CBIR system uses a hierarchical structure by GHSOQM to organize images and GHSOQM must be first trained by using all images. Retrieval processes in some CBIR systems, e.g., SIMPLIcity (Wang et al. 2001), directly compare query image with all images. It uses a flat structure and does not require any training. The extra work by GHSOQM is compensated by a faster retrieval time.

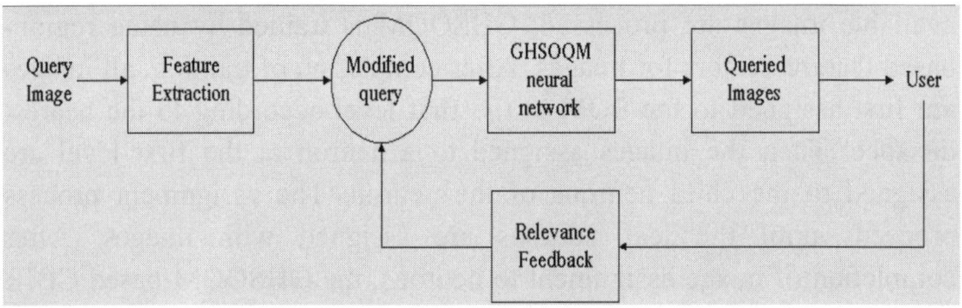

Figure 7.7. Architecture of GHSOQM-based CBIR system

7.2.1.2. Image Segmentation, Feature Extraction and Region-Based Feature Matrices

JSEG algorithm (Deng et al. 1999) is an algorithm for image segmentation. JSEG first quantizes colors in an image and generates a class map. Based on the class map, JSEG finds a good segmentation with coarse or precise resolution by using a criterion for goodness of segmentation.

After image segmentation we can perform feature extraction for each region of an image. Thirteen features are extracted for each region, i.e., six for colors, six for textures and one for region percentages of images. We compute the average and standard deviation of the *L*, *a* and *b* components in *Lab* color space for each region of an image. We denote the average of *L*, *a* and *b* as f_1, f_2, and f_3, the standard deviation of *L*, *a* and b as f_4, f_5, and f_6. For texture features, the following three variables for a 4×4 block in an image as used:

$$a = \sqrt{(\sum_{i=1}^{2}\sum_{j=1}^{2} a_{ij}^2)/4}\ , b = \sqrt{(\sum_{i=1}^{2}\sum_{j=1}^{2} b_{ij}^2)/4}\ , c = \sqrt{(\sum_{i=1}^{2}\sum_{j=1}^{2} c_{ij}^2)/4} \qquad (7.12)$$

where $\begin{bmatrix} a_{11} & a_{12} \\ a_{21} & a_{22} \end{bmatrix}$, $\begin{bmatrix} b_{11} & b_{12} \\ b_{21} & b_{22} \end{bmatrix}$ and $\begin{bmatrix} c_{11} & c_{12} \\ c_{21} & c_{22} \end{bmatrix}$ are the coefficients of Haar wavelet transform for *LH*, *HL* and *HH* band, respectively (Wang et al. 2001). After wavelet transformation, we just assign the three variables to each pixel of the block. Then we compute the average and standard deviation of the three features *a*, *b* and *c* for each region. We denote the average of *a*, *b* and *c* as f_7, f_8 and f_9, the standard deviation of *a*, *b* and *c* as f_{10}, f_{11} and f_{12}. The last feature f_{13} is the region percentage of an image.

So an image *x* can be denoted by region-based features matrix $\begin{bmatrix} R_1^x \\ \dots \\ R_n^x \end{bmatrix}$, where $R_i^x = [f_{i,1}^x, \dots f_{i,12}^x]\,(i=1,\dots,n)$ is a row feature vector representing the *i*th region of the image *x*, *n* is the number of regions in the image. Different images may have different number of regions. For the sake of convenience, a neuron in the GHSOQM-based CBIR system is represented by a feature matrix with a fixed number of rows, which means a fixed number of regions.

7.2.1.3. Image Distance

Since the representation of an image is a feature matrix, we defined a distance measure in order to compare the dissimilarity of two images. In this study, images are compared with direction from query image to other images. The weight assignment for each region of an image is just the region percentage of the image.

Suppose we have two images *A* and *B*. Image *A* have *n* regions and image *B* has *m* ones. The corresponding representing matrix are $\begin{bmatrix} R_1^A \\ \dots \\ R_n^A \end{bmatrix}$

and $\begin{bmatrix} R_1^B \\ \dots \\ R_m^B \end{bmatrix}$, where R_i^A is the row feature vector of the ith region of image A and each component of all feature vectors are normalized to lie in [0 1]. The region distance between R_m^A and R_n^B is defined as the following:

$$d_{mn} = \left(w_1 \sum_{i=1}^{3} (\bar{f}_{mi}^A - \bar{f}_{n,i}^B)^2 + w_2 \sum_{i=4}^{6} (\bar{f}_{mi}^A - \bar{f}_{n,i}^B)^2 + w_3 \sum_{i=7}^{9} (\bar{f}_{mi}^A - \bar{f}_{n,i}^B)^2 + w_4 \sum_{i=10}^{12} (\bar{f}_{mi}^A - \bar{f}_{n,i}^B)^2 \right)^{1/2} \tag{7.13}$$

where w_1 to w_4 are the weights to colors and textures. The weights are chosen such that $w_1 = w_3 = w_4 = 1$, and $w_2 = 0.5$. With this selection of weights for colors and textures, image retrieval results are satisfactory.

The distance from image A to B is described as the following steps.

Step 1) Compute the distance matrix D, where d_{mn} is the element in mth row and nth column of the matrix and denoted by Eq. (7.13).

Step 2) Find the minimum value in each row of the matrix D and denote D_i as the minimum of ith row of D.

Step 3) Compute the weighted average for distance from image A to B:

$$\text{Distance(A,B)} = \sum_{i=1}^{n} \bar{f}_{i,13}^A D_i \tag{7.14}$$

The above distance of two images can be illustrated in Fig. 7.8. First find the nearest regions from image A to B. The nearest regions are connected in Fig. 7.8. Then the distance from image A to B is

$$\text{Distance(A,B)} = f_{1,13}^A d_{13} + f_{2,13}^A d_{22}$$

7.2.1.4. GHSOQM and Relevance Feedback in the CBIR System

The GHSOQM algorithm is used in the CBIR system. Images and weights of neurons are represented by feature matrices. A large number of computations in GHSOQM is to find the nearest neurons to retrieve images. As mentioned before, a neuron i represents an image at time t by $w_i = \begin{bmatrix} R_1^i(\mathrm{t}) \\ \cdots \\ R_r^i(t) \end{bmatrix}$, where r is the fixed number of regions. The distance from an image A to neuron i (with weight matrix w_i) is the same function as Eq. (7.14):

$$\text{Distance(A,)} = \sum_{i=1}^{n} \overline{f}_{i,13}^{A} D_i . \tag{7.15}$$

where D_i is the minimum value in the ith row of the distance matrix D between image A and neuron i.

The weight updating for neurons must be modified in the CBIR system because of the matrix representation of images. The weight updating now is the following steps:

Step 1) Find the nearest regions of an updating neuron k from a query image $x = \begin{bmatrix} R_1^x \\ \cdots \\ R_n^x \end{bmatrix}$ at time t. The found regions of the neuron are arranged with order in a matrix $\begin{bmatrix} R_1^k(t)' \\ \cdot\cdot \\ R_n^k(t)' \end{bmatrix}$ corresponding to the regions of x. Note that the found regions may repeat such that $R_i^k(t)'$ and $R_j^k(t)'$ are the same as a region of the neuron k.

Step 2) Update the neuron by:

$$R_i^k(t+1)' = R_i^k(t)' + \varepsilon(t) h_{kc}(R_i^x - R_i^k(t)'), \quad i = 1,\ldots,n \tag{7.16}$$

where h_{kc} is the neighborhood function , R_i^x is the feature vector of the ith region of the query image x, $R_i^k(t)'$ is the feature vector of the nearest region in neuron k at time t from the ith region of the query image x.

For example, a query image x has three regions $\begin{bmatrix} R_1^x \\ R_2^x \\ R_3^x \end{bmatrix}$ and an updating neuron k have four $\begin{bmatrix} R_1^k \\ R_2^k \\ R_3^k \\ R_4^k \end{bmatrix}$. As shown in Fig. 7.9, the corresponding nearest regions of the neuron from the query image are $\begin{bmatrix} R_2^k \\ R_3^k \\ R_3^k \end{bmatrix}$ with order. The weight updating is

$$\begin{aligned} R_2^k(t+1) &= R_2^k(t) + \varepsilon(t)h_{kc}(R_1^x - R_2^k(t)) \\ R_3^k(t+1) &= R_3^k(t) + \varepsilon(t)h_{kc}(R_2^x - R_3^k(t)) \\ R_3^k(t+1) &= R_3^k(t) + \varepsilon(t)h_{kc}(R_3^x - R_3^k(t)) \end{aligned} \tag{7.17}$$

Note that the region 3 of the neuron k is updated twice because it is the nearest region from the region 1 and 3 of the query image.

Assume the query image has a feature matrix $x = \begin{bmatrix} R_1^x \\ \dots \\ R_n^x \end{bmatrix}$. The retrieved images are classified as relevant images $Y = \{Y_1, \dots, Y_{|Y|}\}$ and irrelevant images $Z = \left\{Z_1, \dots, Z_{|Z|}\right\}$. For image Y_i, find the nearest regions from the regions of the image x and denote them as a matrix $Y_i' = \begin{bmatrix} Y_{i1}' \\ \dots \\ Y_{in}' \end{bmatrix}$, where Y_{ij}' is the nearest region of image Y_i from the jth region of the query image x. Similarly, the nearest regions from the image x are denoted as a matrix $Z_i' = \begin{bmatrix} Z_{i1}' \\ \dots \\ Z_{in}' \end{bmatrix}$ for each Z_i. Then the new query matrix is modified by

$$R_i^x(t+1) = \alpha R_i^x(t) + \frac{\beta}{|Y|}\sum_{Y_k \in Y} Y_{ki}' - \frac{\gamma}{|Z|}\sum_{Z_k \in Z} Z_{ki}', \quad i = 1,\dots,n \qquad (7.18)$$

where α, β and γ are parameters controlling the relative weighting of current query image, relevant images and irrelevant images, respectively.

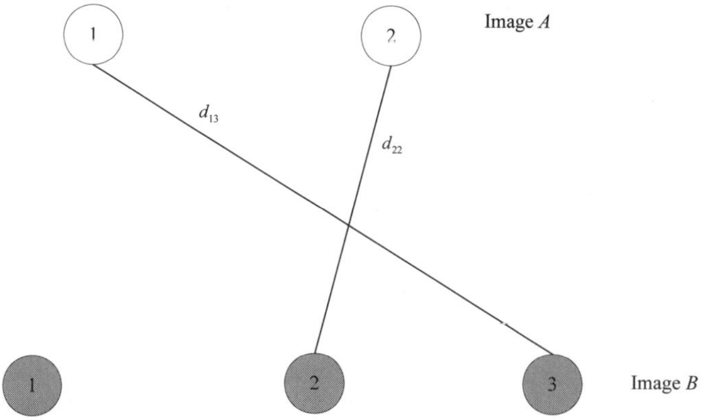

Figure 7.8. Illustration of image distance by the GHSOQM-based CBIR system

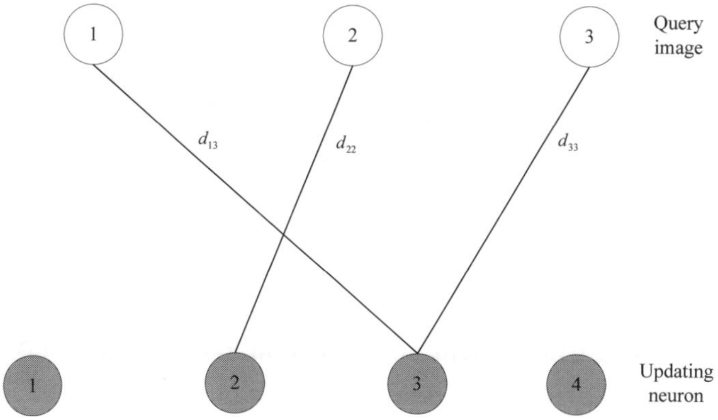

Figure 7.9. Illustration of nearest regions of an updating neuron from a query image

7.2.2. Performance Evaluation

In this experiment a database of one thousand images were used to test the effectiveness of the CBIR system. The number of fixed regions for representing a feature image of one neuron is set to 10. After image retrieval, users select relevant images with several rounds of relevant feedback iterations. The next new query by RF leaves out already selected images. And all selected images are always listed on the top rank.

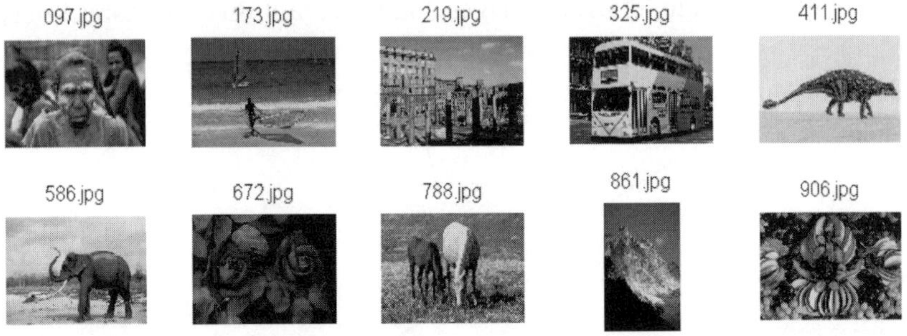

Figure 7.10. Ten sample images with one sample image in each class. The name of an image is shown on the top of the image

In this data set, 1000 images from (Wang et al. 2001) are chosen to test the system. The images have ten classes, each of which contains 100 pictures. The sizes of the images are 384×256 or 256×384. Ten sample images with one sample image in one class are shown in Fig. 7.10. The parameter τ for GHSOQM training was set to 20, which means that a neuron must generate its child neurons if the number of images belonging to it is more than 20. Another parameter λ for the retrieval process was set to 20, which means that the least number of a target neuron for image retrieval is more than 20. Furthermore, the implemented GHSOQM-based CBIR system only shows the first 20 images to users. After completion of training GHSOQM, the CBIR system was ready for testing.

Firstly the CBIR system requires training for GHSOQM. The average training time by GHSOQM is 1890 seconds for the 1000 images. The querying time of the GHSOQM-based is faster than that of direct query method that compares a query image with all other images of the database. The average search time of GHSOQM-based query for one image is 3.08 seconds and direct query is 7.25. As shown in Fig. 7.11, the proposed query is much faster than direct query because the hierarchical structure of images and the additional training before querying.

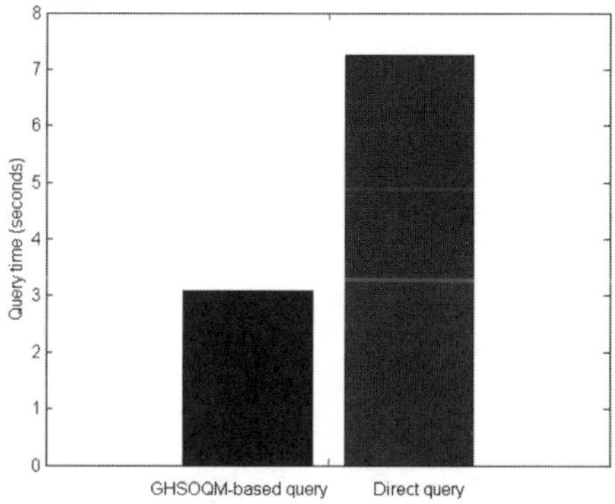

Figure 7.11. Comparison of query time between GHSOQM-based query and direct query

The recall-precision graph is used to evaluate the retrieval results. Precision P is defined as the following:

$$P(k) = n_k / k \tag{7.19}$$

where k is the number of retrieved images and n_k is the number of relevant images in the retrieved images. Recall R is defined as

$$R(k) = n_k / N \tag{7.20}$$

where N is the number of all relevant images in the data set. An optimal recall-precision graph would have a straight line, i.e., precision always at 1. Typically, when recall increases, precision decreases. Since the system only shows the first 20 images, the maximum value of recall is 0.2. The aforesaid 10 sample images from all classes are used and tested the performance of GHSOQM-based query and direct query. The recall-precision graphs are plotted in Fig. 7.12. GHSOQM-based query and direct query have similar query performance. The recall-precision graphs of some images are optimal at the recall interval [0 0.2], i.e., buses, dinosaurs and flowers. This is because the objects in these classes have simpler color distributions. The performance for the images from other classes is degraded because the objects in the images have more complex color distributions.

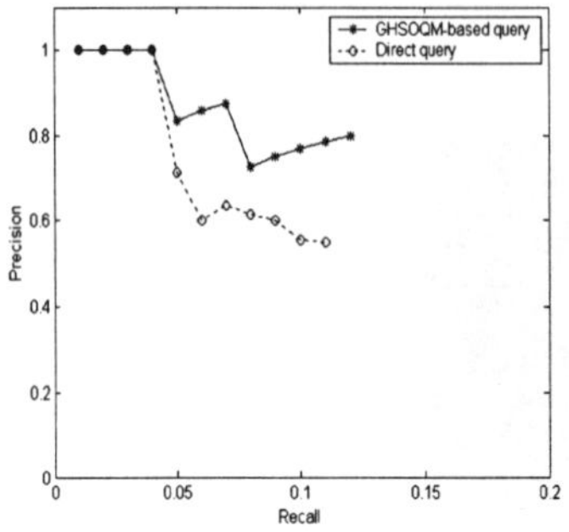

(a)

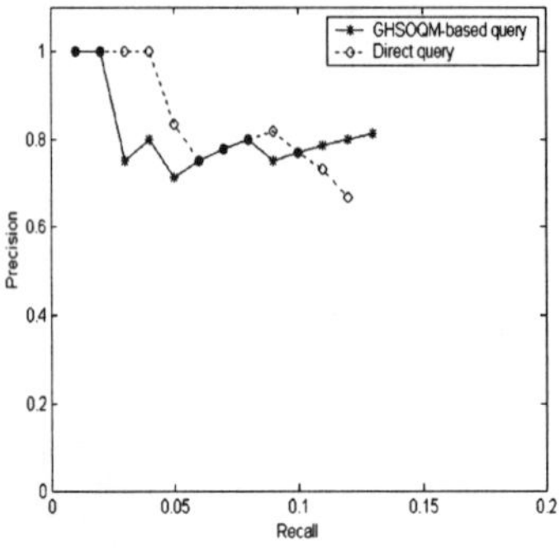

(b)

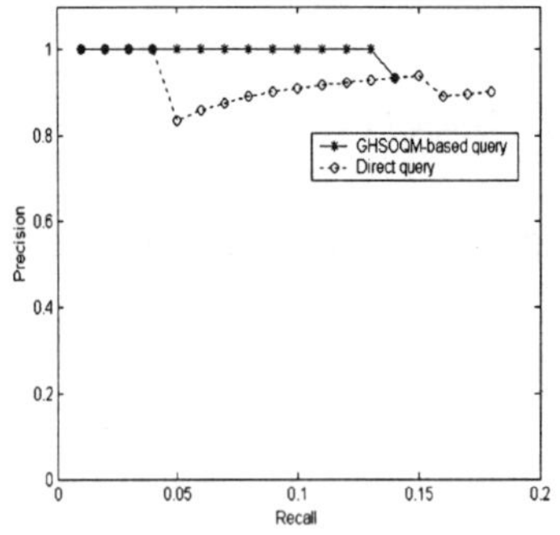

(c)

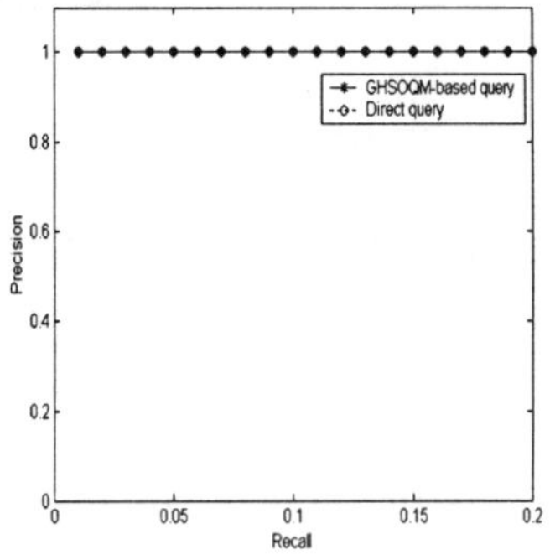

(d)

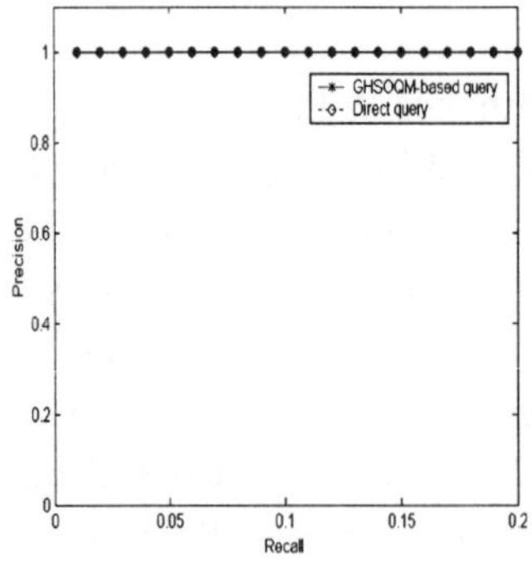

(e)

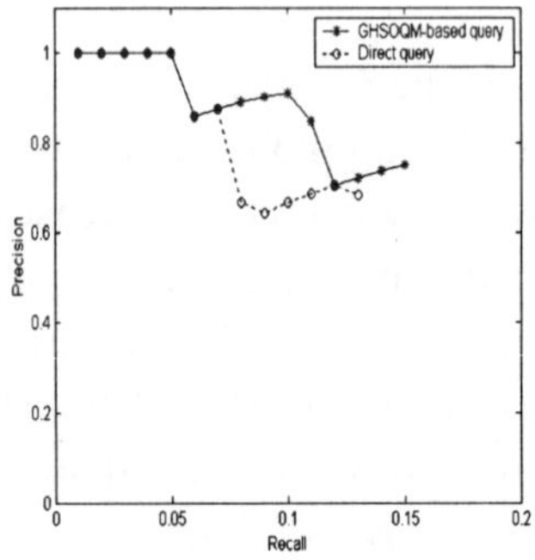

(f)

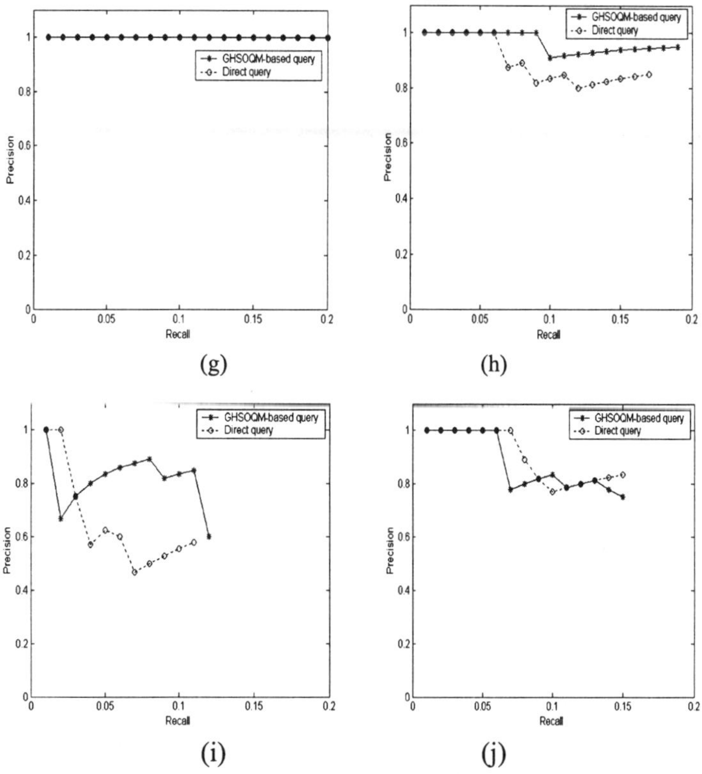

Figure 7.12. Recall-precision graphs for GHSOQM-based query and direct query on the 10 images (a) 097.jpg (African people and village) (b) 173.jpg (beach) (c) 219.jpg (building) (d) 325.jpg (buses) (e) 411.jpg (dinosaurs) (f) 586.jpg (elephants) (g) 672.jpg (flowers) (h) 788.jpg (horses) (i) 861.jpg (mountains and glaciers) (j) 906.jpg (food)

7.3. Feature Selection for cDNA Microarray

In this section, we describe how feature selection technique described in Chapter 6 be applied to bioinformatics problem. Microarrays are a powerful biotechnological means because they are able to record the expression levels of thousands of genes simultaneously. Through hybridizing the fluorescent DNA probe of an examined sample with that of a reference cell, the mRNA levels of the genes in the examined sample are obtained. Since the mRNA levels are roughly related to the amount of protein product, the obtained microarray result can be used to express the "state" of the examined sample. Generally, different cells or a cell under different conditions yield different microarray results. The comparisons of microarray results between normal and cancer cells can provide the important information of cancer diagnosis and treatment. Among a large amount of genes encoded in the microarray gene expression data, only a very small fraction of them are informative for a certain task. A very challenging task arises as a result – how to select the most useful features (genes) for performing data analysis such as diagnosis, prognosis, subtype classification of a heterogeneous disease and understanding of a gene network. The gene selection is important and sometimes necessary because of two main reasons. First, it is impossible for biologists or physicians to examine the whole feature space (e.g. the genes in human genome) in the laboratory experiments at one time. It is necessary to recommend a small fraction of the features by using computational algorithms. Second, it is widely known that taking many irrelevant features into account amid the course of classification will increase the dimensionality of the problem, and thus results an unnecessary computational difficulties and additional noise.

In this section, we describe combining the quadratic MI based feature selection method with a grid based clustering algorithm, called QMIFS-GC in short in this section, for performing gene selection. The method consists of two sequential parts:

1) A supervised QMIFS-GC algorithm is used to sort out and discard the highly redundant features. As a result, the computational efficiency of the whole feature selection process is greatly improved without reducing the quality of the selection results.

2) In the MI based forward selection stage, the quadratic MI estimation and Gaussian based probability estimators is employed. With them, MI can be estimated effectively even when only a small pattern set is available. Also, we can use a MI based criterion to filter out the redundant features. Finally, the direct MI estimation enables us to terminate the selection process at an appropriate point where the selected gene subset has preserved the most essential information of the given microarrays dataset.

7.3.1. MI Based Forward Feature Selection Scheme

The forward searching strategy described in Chapter 6 is used because of its simple implementation and its relative high efficiency. First let us consider the microarrays problem as a general classification dataset $\{X; C\}$, in which the feature set is denoted by F, a MI based forward feature selection algorithm is generally realised as follows.

(Initialization) Set the selected feature set S empty.

For any feature (say, f_i) in F, compute the MI $I(f_i ;C)$.

Determine the feature that maximizes $I(f_i ;C)$. Add hat feature into S, and delete it from F.

Repeat the following two steps until stopping criterion is met

Calculate $I(S+f_i ;C)$, for any feature (say, f_i) remaining in F.

Choose the feature that maximizes $I(S+f_i ;C)$. And put that feature into S, and eliminate it from F.

Output the selected feature subset S.

In the above processes, the MI between S and the output variable C increases gradually because the adding of input variables cannot decrease MI. The incremental MI gradually decreases to zero when all the relevant genes (features) are selected. Assume that f_a is a selected gene at certain iteration (say ith iteration), S is the selected gene subset before this iteration. The incremental MI at ith iteration is the conditional MI $I(f_a;C|S)$. Assume that f_b is the next selected gene, the incremental MI of the $(i+1)$th iteration is $I(f_b;C|S+f_a)$. In the above process,

$$I(f_a;C\,|\,S) > I(f_b;C\,|\,S) \tag{7.21}$$

Based on the definition of the conditional MI, we have

$$I(f_b;C\,|\,S) = I(f_b;C\,|\,S,f_a) + I(f_a;C\,|\,S) \geq I(f_b;C\,|\,S,f_a) \tag{7.22}$$

With Eqs. (7.21) and (7.22), we have $I(f_a;C|S) > I(f_b;C|S+f_a)$. This inequality suggests that the incremental MI decreases in the above searching process. As a result, the forward process can be reliably terminated when the incremental MI is small enough implying the unselected genes at that point contain little additional information for cancer diagnosis. The small sample set and huge gene set of a microarray gene expression data poses two main challenges to the above MI based feature selection scheme. First, a relatively small sample set makes the estimation of high-dimensional MIs much harder. Second, a large amount of genes leads to a remarkably huge computational burden. These difficulties must be addressed.

7.3.2. The Supervised Grid Based Redundancy Elimination

In a microarray dataset, many redundant genes exist. Filtering out those redundant genes efficiently before performing feature selection will greatly enhance the computational efficiency. Simply using other conventional gene clustering methods to reduce the redundancy is very computationally demanding. Alternatively, based on the concept of grid, a simple and fast algorithm can identify redundant genes in an efficient way. In details, the basic concept of the grid-based redundancy elimination algorithm is that objects in a grid must be similar to each other when the size of that grid is small enough. Due to the sparsity in the high-dimensional spaces, the size of the grid becomes critical. In order to enhance the performance of robustness, an adaptive grid size, rather than a fixed grid size, is used. Using the property of MI ranking, only the features with close MI values are checked if they are within the grid. For a considered feature *f*, if a feature has similar MI value to *f* and falls within the grid around *f*, it will be removed as a redundant feature.

At the beginning of the clustering process, the MI of each gene with output variable is estimated as described in the last chapter. With these estimates, the discrimination abilities of genes are evaluated. Clustering is performed on each gene in a descending order of the MI estimates unless the ones are marked as redundant one. In each iteration, an adaptive grid is generated around the considered gene and only genes with the acceptable MI values are checked. The grid size starts at the maximum distance different in a dimension and changes until the number of redundant genes is within the pre-defined range, *GridNumRange*. The number of genes is defined by the user or determined according to the MI estimate differences. There are two types of input parameters: 1) the number of genes within the grid, *GridNumRange*, and 2) the number of genes for checking redundancy, *RedNum*. The number of genes within the grid should be given to guide the changing of the grid. The number of genes for checking redundancy could be defined by fixing the number of genes directly or input the acceptable MI difference, which is used to determine the number of genes for redundancy check. The changing of the grid size depends upon the number of redundant genes. This QMIFS-GC algorithm is realized as follows.

7.3.3. The Forward Gene Selection Process Using MIIO and MISF

As illustrated in Fig. 7.13, the gene selection method consists of two sequential processes – the supervised QMIFS-GC process, and the MI based forward feature selection process. Suppose that *R* is the result of the grid-based redundancy elimination. In the MI based forward process, the genes (features) in *R* are firstly ranked in a descend order of *MIIO*. The gene satisfying two constraints – having as the large *MIIO* as possible and not being redundant to the selected gene subset (determined by using *MISF*) – is identified and placed into the selected gene subset *S*. This process repeats until no unselected important gene is found. Using *R*, the forward selection process can be stated as follows.

Step 1. R is the result of the above clustering process. And the selected gene set (*S*) is set empty.
Step 2. Calculate *MIIO(f)* for each gene *f* in R. According to *MIIO(f)*, sort out the most important gene, f_k. Put f_k into S, delete f_k from R, and set $MIIO_1 = MIIO(f_k)$.
Step 3. Estimate $MIIO(S + f)$ for each gene *f* remaining in R.
Step 4. Identify f_k. having $MIIO(S + f_k) = \arg\max_i(MIIO(S + f_i))$, and delete f_k. from *R*.
Step 5. If the candidate feature f_k is not redundant to *S*, i.e., $MISF(f_k; S) \leq 0.9$, put f_k into *S*, set $MIIO_j = MIIO(S)$ (*j* is the number of the features in *S*), otherwise, goto Step 4.
Step 6. If $(MIIO_j - MIIO_{j-1}) / MIIO_1 \leq \gamma$, goto Step 7, otherwise, goto Step 3.
Step 7. Output the gene subset *S*.

The threshold in the stopping criterion γ is set with 0.05. With $\gamma = 0.05$, we know that the information beyond the selected gene subset is small enough to be ignored.

7.3.4. Results

In this section, we illustrate how the quadratic MI combining with a grid based clustering method be used for selecting genes. This method is also compared with other two selection methods, one is called FR for short in this section, and another one is called the SVM RFE. Assuming that all genes are independent to each other, FR ranks genes according to the individual linear discriminant ability. To rank the genes, SVM RFE depends on SVM, a state-of-art classification model: SVM RFE firstly builds a linear SVM model using all the genes, and then according to the parameters of the built SVM model it ranks genes in a descending order of classification importance. Through discarding low-ranked ones, the current gene set is reduced by half. The process of building-SVM-discarding-half-of-genes repeats until no gene remains. We also demonstrate the QMI based feature forward selection method (QMIFS), which conducts the forward feature selection on the whole gene set. There is no grid-based clustering for pre-processing. Four different types of classifiers are used to evaluate the gene selection results. They are two types of support vector machine models (SVM), decision tree (DT) and *k*-NN rule. Decision tree and *k*-NN rule are available in the Weka software package (at http://www.cs.waikato.ac.nz/~ml/weka). Following Guyon, we downloaded SVM model from http://www.isis.ecs.soton.ac.uk /resources/svminfo, and used two types of SVM models – the linear SVM model (SVM-L) and the RBF SVM model (SVM-R). When all these classifiers achieve the best or the near-best performance before the stopping points, it is assumed that the selection results have covered most of the important information.

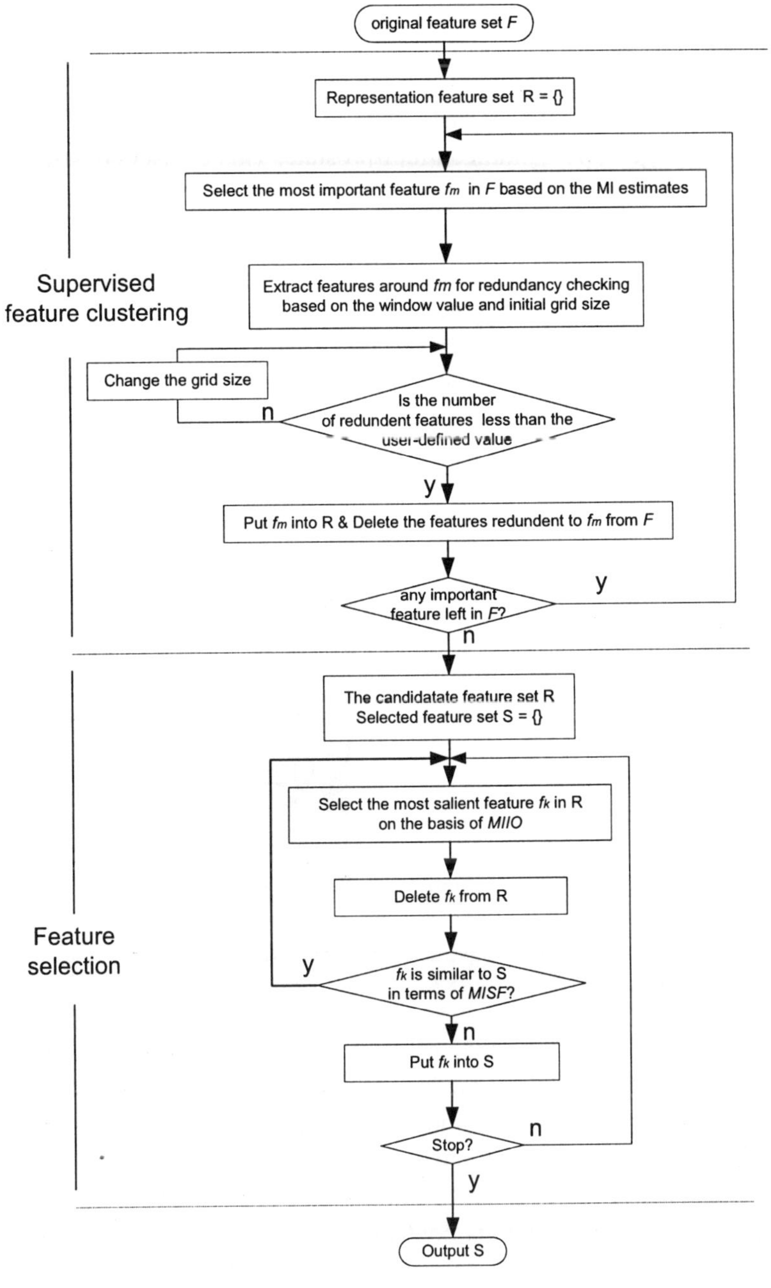

Figure 7.13. The block diagram of the QMIFS-GC method

7.3.4.1. Prostate Cancer Classification Dataset

The objective of this task is to distinguish prostate cancer cases from non-cancer cases. The original raw data are published at http://www.genome.wi.mit.edu/mpr/prostate. This dataset consists of 102 samples from the same experimental conditions. And each sample is described by using 12600 features. We split the 102 samples into two disjoint groups – one group with 60 samples for training and the other one with 42 samples for testing.

First, QMIFS and the grids based MI method select the best 50 genes. These selected genes are compared with those obtained from FR and SVM RFE in terms of efficiency and effectiveness as shown in Fig. 7.14 and Table 7.3 respectively.

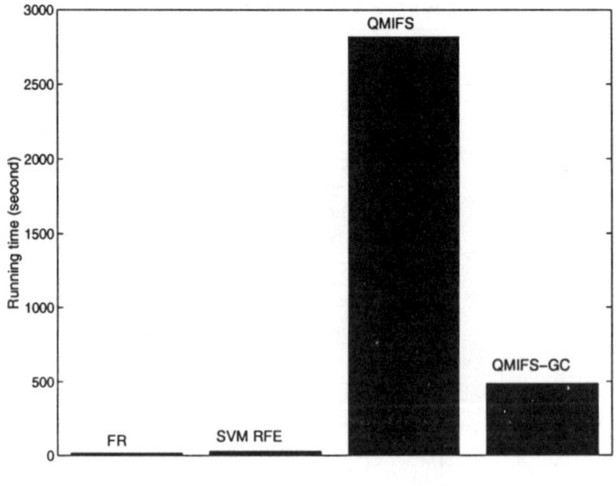

Figure 7.14. Comparisons in terms of the running time on the prostate cancer classification data

These results show that FR and SVM RFE are much faster than QMIFS and grid-based MI method because the searching strategies in FR and SVM-RFE are very simple – FR only ranks features individually, and SVM RFE reduces the remaining features in an exponential rate. The comparisons between QMIFS and the QMIFS-GC clearly suggest the huge computational savings contributed by the redundancy elimination approach. In practice, this process could reduce the number of genes from 12,600 to 872 with less than 4 minutes. The results listed in Table

7.3 indicate that QMIFS and QMIFS-GC have very similar gene selection effectiveness. These two methods provide better results than the SVM RFE and FR methods. With the (near) best effectiveness and the better efficiency, the QMIFS-GC delivers good gene selection results.

Table 7.3 Comparisons of classification accuracy on the prostate cancer classification dataset

Number of selected features		FR	SVM-RFE	QMIFS	QMIFS-GC
k-NN	4	0.83	0.88	0.93	0.93
	8	0.83	0.93	0.95	0.95
	16	0.83	0.90	0.88	0.90
SVM-R	4	0.78	0.90	0.95	0.95
	8	0.81	0.93	0.95	0.95
	16	0.86	0.93	0.93	0.95
SVM-L	4	0.76	0.93	0.95	0.95
	8	0.83	0.93	0.98	0.98
	16	0.83	0.90	0.98	0.95
Decision Tree	4	0.76	0.88	0.88	0.88
	8	0.76	0.81	0.90	0.90
	16	0.71	0.81	0.90	0.90

In Fig. 7.15, the changes of *MIIO* and the incremental *MIIO* are illustrated. They imply that the gene selection process stops when 25 genes are selected. And all classifiers are able to deliver their best or near-best performance before the stopping point, as illustrated in Fig. 7.16. In Table 7.4, the top 8 genes selected are briefly described. Each gene basically carries different biological meaning and exhibits different biological function. For example, 37639_at, which is also determined as one of the genes for prostate cancer classification, is for human hepatoma mRNA for serine protease and it plays an essential role in cell growth and maintenance of cell morphology (referred to http://www.rzpd.de/cgi-bin/cards/). Further details on these genes can be found in the websites about genomics, such as, http://expression.gnf.org/cgi-bin/index.cgi.

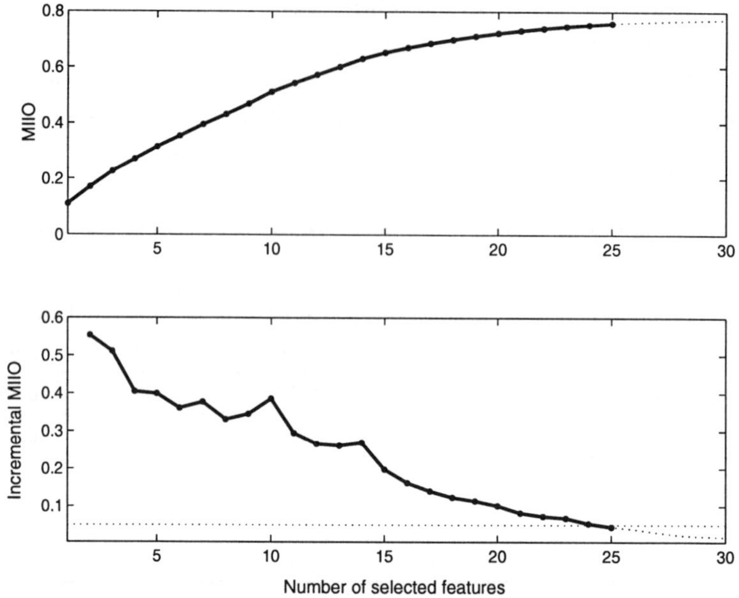

Figure 7.15. The change of MIIO and the incremental MIIO with the number of the selected genes on the prostate classification data

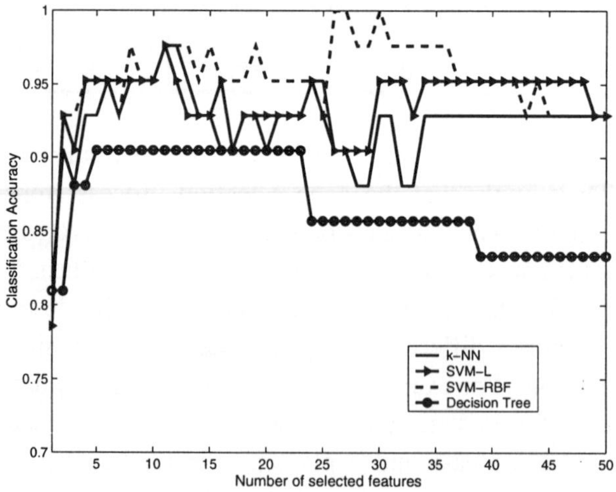

Figure 7.16. Classification results of the genes selected by QMIFS-GC on the prostate cancer classification data

Table 7.4 The gene selection result of QMIFS-GC on the prostate cancer classification dataset. GAN represents the gene accession number

Selection Order	Reference Number	GeneBank	Description
1	37639_at	X07732	Human hepatoma mRNA for serine protease hepsin
2	37720_at	M22382	Human mitochondrial matrix protein P1 (nuclear encoded) mRNA.
3	38028_at	AL050152	Homo sapiens mRNA; cDNA DKFZp586K1220 (from clone DKFZp586K1220)
4	41504_s_at	AF055376	Homo sapiens short form transcription factor C-MAF (c-maf) mRNA.
5	32786_at	X51345	Human jun-B mRNA for JUN-B protein
6	36864_at	AJ001625	Homo sapiens mRNA for Pex3 protein
7	35644_at	AB014598	Homo sapiens mRNA for KIAA0698 protein.
8	38087_s_at	W72186	zd69b10.s1 Homo sapiens cDNA.

7.3.4.2. Subtype of ALL Classification Dataset

The pediatric acute lymphoblastic leukemia (ALL) is a heterogeneous disease. The correct diagnosis of the subtypes for a patient is crucial because different subtypes have different treatment plan. Over-treated or less-treated therapy could lead to serious consequences to the patient. The subtype classification of this disease has been comprehensively studied previously using gene expression profiling and supervised machine learning methods. The original data is divided into six diagnostic groups (BCR-ABL, E2A-PBX1, Hyperdiploid>50, MLL, T-ALL and TEL-AML1), and a miscellaneous class that contains diagnostic samples that did not fit into any one of the above groups (thus labeled as "Others"). There are total of 12558 features and 327 samples in this dataset. This dataset is partitioned into two disjoint subsets, in which 215 samples are used for training and 112 are used for testing.

Comparative results are shown in Fig. 7.17 and Table 7.5. The running time shown in Fig. 7.17 is the time required for selecting 150 features. The change of *MIIO* is shown in Fig. 7.18, which shows the gene selection process stops when 95 genes are selected. In Fig. 7.19, it indicates that the best or the near best classification results could be obtained before this stopping point. Also, by using the classification schemes, the results are compared with other reported results. These results are summarized in Table 7.6. In Table 7.7, the top 20 selected genes are listed.

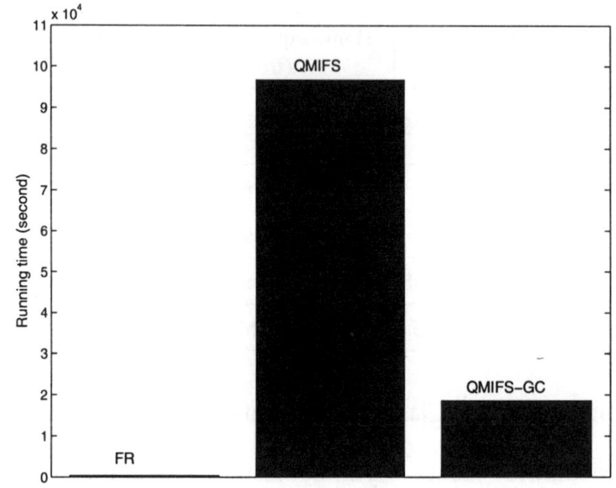

Figure 7.17. Comparisons in terms of running time on the ALL subtype classification data

Table 7.5 Comparisons of classification accuracy on the ALL subtype classification dataset

Number of selected features		FR	QMIFS	QMIFS-GC
k-NN	4	0.46	0.57	0.57
	8	0.74	0.79	0.79
	16	0.73	0.88	0.88
	32	0.72	0.89	0.90
Decision Tree	4	0.43	0.58	0.58
	8	0.71	0.76	0.76
	16	0.72	0.79	0.79
	32	0.76	0.79	0.79

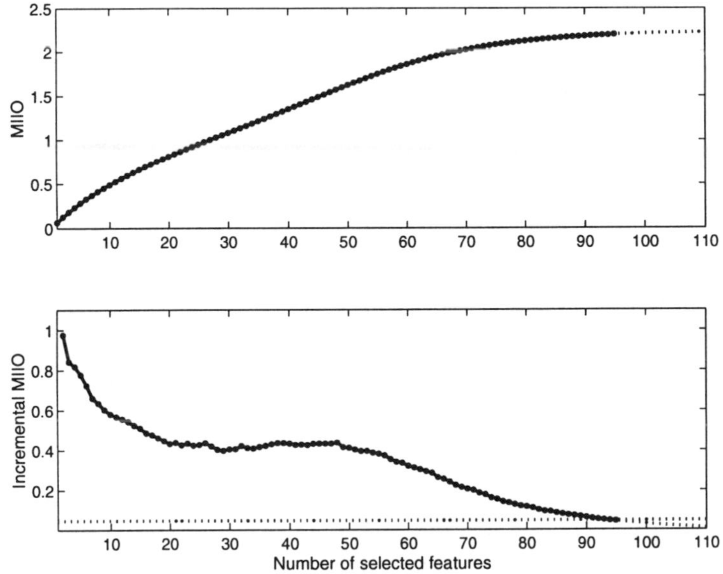

Figure 7.18. The change of *MIIO* and the incremental *MIIO* with the number of the selected genes on the ALL subtype classification data

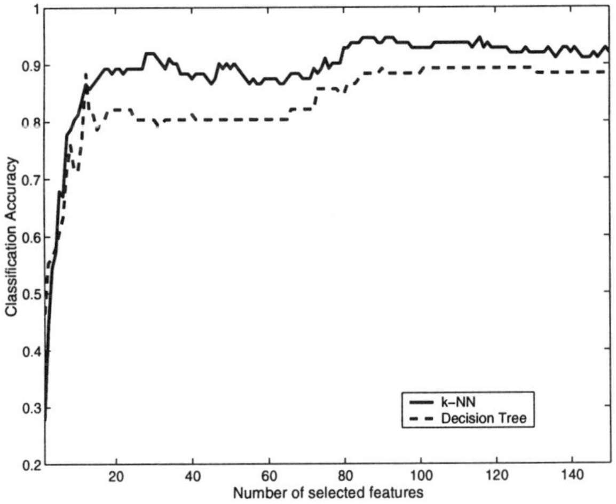

Figure 7.19. Classification results of the genes selected by QMIFS-GC on the ALL subtype classification data

Table 7.6 Comparisons between QMIFS-GC and the other methods. The numbers listed in this table are classification accuracy. To evaluate a gene selection method, binary classification schemes for each subtype are constructed by using 50 genes determined by that gene selection method. The details about this classification scheme can be found in [6] (a) The results of SVM (b) The results of *k*-NN rule

(a)

	QMIFS-GC	Chi sq	CFS	T-stats	SOM/DAV
T-ALL	1.00	1.00	1.00	1.00	1.00
E2A-PBX1	1.00	1.00	1.00	1.00	1.00
TEL-AML1	0.98	0.99	0.99	0.98	0.97
BCR-ABL	0.97	0.95	0.97	0.94	0.97
MLL	0.96	1.00	0.98	1.00	0.97
H>50	0.97	0.96	0.96	0.96	0.95

(b)

	QMIFS-GC	Chi sq	CFS	T-stats	SOM/DAV
T-ALL	1.00	1.00	1.00	1.00	1.00
E2A-PBX1	1.00	1.00	1.00	1.00	1.00
TEL-AML1	0.98	0.98	0.98	0.99	1.00
BCR-ABL	0.97	0.94	0.97	0.95	0.93
MLL	0.96	1.00	0.98	0.95	1.00
H>50	0.97	0.98	0.96	0.94	0.98

Table 7.7 The gene selection results of QMIFS-GC on the ALL subtype classification dataset. GAN represents the gene accession number

Selection Order	Reference Number	GAN	Description
1	1077_at	M29474	Human recombination activating protein (RAG-1) gene.
2	36239_at	Z49194	H.sapiens mRNA for oct-binding factor.
3	41442_at	AB010419	Homo sapiens mRNA for MTG8-related protein MTG16a.
4	38319_at	AA919102	Homo sapiens cDNA.
5	36937_s_at	U90878	Homo sapiens carboxyl terminal LIM domain protein (CLIM1) mRNA.
6	35614_at	AB012124	Homo sapiens TCFL5 mRNA for transcription factor-like 5.
7	38968_at	AB005047	Homo sapiens mRNA for SH3 binding protein.
8	36985_at	X17025	Human homolog of yeast IPP isomerase.
9	38518_at	Y18004	Homo sapiens mRNA for SCML2 protein.
10	41097_at,	AF002999	Homo sapiens TTAGGG repeat binding factor 2.
11	33355_at	AL049381	Homo sapiens mRNA; cDNA DKFZp586J2118 (from clone DKFZp586J2118).
12	38596_i_at	D50402	Human mRNA for NRAMP1.
13	36620_at	X02317	Human mRNA for Cu/Zn superoxide dismutase
14	38242_at	AF068180	Homo sapiens B cell linker protein BLNK mRNA, alternatively spliced.
15	39728_at	J03909	Human gamma-interferon-inducible protein (IP-30) mRNA.
16	38652_at	AF070644	Homo sapiens clone 24742 mRNA sequence.
17	39878_at	AI524125	Homo sapiens cDNA.
18	2087_s_at	D21254	Human mRNA for OB-cadherin-1.
19	37344_at	X62744	Human RING6 mRNA for HLA class II alpha chain-like product.
20	35974_at	U10485	Human lymphoid-restricted membrane protein (Jaw1) mRNA.

Table 7.8 Comparisons on other cancer classification problems. In the columns for listing best classification accuracy, the left value is the best classification accuracy of top 50 feature subsets, and the right value is the smallest size of the gene subsets with the best classification accuracy. The running time of QMIFS and QMIFS-GC is the time for them selecting 50 genes

Feature selection methodology	Running time (second)	Best classification accuracy			
		k-NN	SVM-R	SVM-L	DT
ovarian cancer classification					
FR	46	0.99; 26	0.98; 9	1.00; 22	0.96;10
SVM RFE	351	1.00; 8	1.00; 4	1.00; 4	0.96; 4
QMIFS	2.0×10^4	1.00; 3	1.00; 3	1.00; 3	0.99; 3
QMIFS-GC	1.3×10^3	1.00; 3	1.00; 3	1.00; 3	0.99; 3
colon cancer classification					
FR	1.2	0.76; 3	0.86; 4	0.90; 12	0.81; 8
SVM RFE	3.5	0.86; 8	0.90; 8	0.81; 8	0.76; 8
QMIFS	211.6	0.86; 3	0.90; 9	0.95; 9	0.81; 3
QMIFS-GC	81.1	0.90; 11	0.95; 11	0.90; 3	0.81; 3

Also, we show how these gene selection methods are applied to other microarray type data, such as the colon cancer classification data and the ovarian cancer classification data (the proteomic data of this application were treated in the same way with the microarray data). In the ovarian cancer classification, there are 253 data samples and 15154 genes. Among these samples, 91 are control samples (non-cancer) while 162 are cancer samples. We randomly select 150 samples for training, and the others for testing. The colon cancer classification dataset consists of 62 samples and 2000 genes. The 62 samples are randomly split into two disjoint parts – one part of 40 samples for training and the other of 22 samples for testing. The results are summarized in Table 7.8.

7.3.5. Remarks

We demonstrate how mutual information based feature selection scheme be used for gene selection. It is interesting to show that mutual information is employed for three purposes. First, with the guidance of mutual information, we have demonstrated that the QMIFS-GC can greatly eliminate the redundancy in a huge feature set. Certainly, this concept can be used to other feature selection applications. As a result of the grid based clustering approach, it is worth noting that the efficiency of the whole feature selection can be enhanced. Second, based on mutual information, the salient features are identified gradually. The computational difficulty of estimating the high dimensional MI is solved. Also, attributed to the characteristics of mutual information, the termination of the searching process is not determined in an ad hoc basis. This is useful to most applications. Third, using mutual information, the highly redundant selection results can be avoided in a systematic way. In this application it is of particular important because of the size of the original dataset. In most other physical applications, similar problems may also be experienced. This is a very useful feature to one who is working on feature selection.

Bibliography

Aarts, E. and Korst J. (1989), Simulated Annealing and Boltzmann Machines: A Stochastic Approach to Combinatorial Optimization and Neural Computing, John Wiley & Sons.

Abu-Mostafa, Y. S. (1995), "Hints", Neural Computation, vol. 7, pp. 639-671.

Alahakoon, D., Halgamuge, S. K., and Srinivasan, B. (2000). Dynamic self-organizing maps with controlled growth for knowledge discovery. IEEE Trans. Neural Networks, 11(3), 601-614.

Ani, M. and Deriche. (2001). An optimal feature selection technique using the concept of mutual information. In Proceedings of A.A ISSPA, Malaysia.

Ash, T. (1989), "Dynamic Node Creation in Backpropagation Neural Networks", Connection Science, vol. 1, No. 4, pp. 365-375.

Baba, N. (1989), "A New approach for Finding the Global Minimum of Error Function of Neural Networks", Neural Networks, vol. 2, pp. 367-373.

Battiti, R. (1992), "First- and Second-order Methods for Learning: between Steepest Descent and Newton's Method," Neural Computation, Vol. 4, No. 2, pp. 144-166.

Battiti, R. (1994), "Using Mutual Information for Selecting Features in Supervised Neural Net Learning", IEEE Trans. Neural Networks, vol. 5, no. 4, July.

Battiti, R. and Tecchiolli G. (1995), "Training Neural Nets with the Reactive Tabu Search", IEEE Transactions on Neural Networks, vol. 6, no. 5, pp. 1185-1200.

Battiti, R. (1994). Using mutual information for selecting features in supervised neural net learning. IEEE Tran. on Neural network, vol. 5, No 4, 537-550.

Bauber, H. -U., and Pawelzik, K. (1992). Quantifying the neighborhood preservation of self-organizing feature maps. IEEE Trans. Neural Networks, 3(4), 570-579.

Bazaraa, M. S., Sherali H. D., and Shetty C. M. (1993), Nonlinear Programming: Theory and Algorithms, 2nd edition, New York, Wiley, 1993.

Berenji, H. R. (1992) Fuzzy logic Controllers. In R.R. Yager and L.A. Zadeh, editors, An Introduction to Fuzzy Logic Applications in Intelligent Systems. Kluver Academic Publishers.

Bianchini, M., Frasconi P. and Gori M. (1995) Learning without local minima in radial basis function networks. IEEE Transactions on Neural Networks, 6(3), pp. 749-756.

Bishop, C. M. (1995), Neural Networks for Pattern Recognition, Oxford University Press Inc., New York.

Blackmore, J., and Miikkulainen, R. (1995). Visualizing high-dimensional structure with the incremental grid growing neural network," Proc. 12th Int. Conf. Machine learning, 55-63.

Blake, C. L., and Merz, C. J. (1998). UCI repository of machine learning databases.

Blanzier, E. and Giordana A. (1995) Mapping symbolic knowledge into locally receptive field networks. In M. Gori and G. Soda, editors, Topics in Artificial Intelligence, vol. 992 of Lectures Notes in Artificial Intelligence, pp. 267-278. Springer-Verlag.

Block, H. D. (1962) The perceptron: a model for brain functioning I. Reviews of Modern Physics 34, pp. 123-135. Reprinted in Anderson & Rosenfeld (1988).

Bonnlander, B. and Weigend, A.S. (1994). Selecting input variables using mutual information and nonparametric density estimation. In Proceedings of International Symposium on Artificial Neural network, Taiwan, 42-50.

Broomhead, D. S. and Lowe D. (1988) Multivariable functional interpolation and adaptive networks. Complex Systems, 2, pp. 321-355.

Broomhead, D. S. and Lowe D. (1988) Multivariable functional interpolation and adaptive networks, Complex Systems 2, pp. 321-355.

Chapelle, O., Vapnik, V., Bousquet, O., and Mukherjee, S. (2002). Choosing multiple parameters for support vector machines. Machine Learning, 46(1), 131-159.

Chen, T. and Chen H. (1995) Approximation capability to functions of several variable nonlinear functions and operators by radial basis function neural networks. IEEE Transactions on Neural Networks, 6(4), pp. 904-910.

Cheng, Y. (1997). Convergence and ordering of Kohonen's batch map. Neural Computation, 9, 1667-1676.

Cho, S.Y. and Chow, T.W.S. (1999), "Training Multilayer Neural Networks Using Fast Global Learning Algorithm – Least Squares and Penalized Optimization Methods", Neurocomputing, vol. 25, no. 1-3, pp. 115-131.

Cho, S. -B. (1997). Self-organizing map with dynamical node splitting: application to handwritten digit recognition. Neural Computation, 9(6), 1345-1355.

Chow, T. W. S., and Wu, Sitao. (2002). Piecewise linear projection based on self-organizing map. Neural Processing Letters, 16, 151-163.

Chow, T. W. S, and Wu, Sitao. (2003). An online cellular probabilistic self-organizing map for static and dynamic data sets. IEEE trans. Circuit and System I, 51(4),732-747.

Chow, T. W. S., and Wu, Sitao. (2004). Cell-splitting grid: a self-creating and self-organizing neural network. Neurocomputing, 57, 373-387.

Chow, T.W.S., Rahman, M.K.M., Wu, S. (2006) Content Based Image Retrieval using Tree-Structured Features and Multi-Layer SOM, Pattern Analysis and Applications, in press.

Chow, T. W. S, Leung, C. T, Neural Network based short-term load forecasting using weather compensation, IEEE Trans on Power system, Vol 11, No. 4, pp.1736-1742, Nov 1996.

Chow, T. W. S, Huang, D., Estimating Optimal Features Subset Using Efficient Estimate of High Dimensional Mutual Information, IEEE Trans on Neural Networks, Vol. 16, No. 1, January 2005, pp.213-224.

Chow, T. W. S, Rahman, M. K. M, Face Matching in Large Database by Self-Organising Maps, Neural Processing Letters, Volume 23, Issue 3, Jun 2006, Pages 305 – 323.

Chow, T. W. S., Rahman, M. K. M., Wu, Sitao, Content Based Image Retrieval by Using Tree-Structured Features and Multi-Layer SOM, Pattern Analysis and Applications 9, May. 2006, 1-20.

Chow, T. W. S, Sitao, An Online Cellular Probabilistic Self-Organizing Map for Static and Dynamical Data Sets, IEEE Trans on Circuit and Systems, Part I, Vol. 51, No. 4, April 2004, pp. 732-747.

Chow, T. W. S, Wu, Sitao, Cell-Splitting Grid: A self-creating and self-organizing Neural Networks, Neurocomputing, Vol.57, Mar. 2004, pp. 373-387.

Chow, T.W.S, Yong, Fang, A recurrent Neural Network Based Real-Time Learning Control Strategy Applying to Nonlinear Systems with Unknown Dynamics, IEEE Trans On Industrial Electronics, Vol. 45, No. 1, pp.151-161, Feb, 1998.

Cloete, I. Zurada J. M. (2000) Knowledge-Based Neurocomputing. MIT press, Boston.

Cortes, C. Vapnik V. (1995) Support Vector Networks, Machine Learning, 20(3), pp. 273-297.

Cover, T. M. and Thomas, J. A. (1991). Elements of Information Theory. New York. John Wiley.

Cover, T.M. and Thomas, J.A. (1994). Elements of information theory. New York: John Wiley.

Das, S.K., (1971). Feature selection with a linear dependence measure. IEEE. Trans. Computers, 1106-1109.

Dash, M. and Liu, H. (1997). Feature selection for classification. Intelligent data analysis, 131-156.

De Boor, C. (1978) A Practical Guide to Splines. New York: Springer.

Deco, G., Finnoff W. and Zimmermann H. G. (1995), "Unsupervised Mutual Information Criterion for Elimination of Overtraining in Supervised Multilayer Networks," Neural Computation, Vol. 7, pp. 86-107.

Demartines, P. and Hérault, J. (1997). Curvilinear component analysis: a self-organzing neural network for nonlinear mapping of data sets. IEEE trans Neural Networks, 8(1), 148-154.

Dempster, A., Laird, N., and Rubin, D. (1977). Maximum likelihood from incomplete data via the EM algorithm. Journal of the Royal Statistical Society B, 39, 1-38.

Denoeux, T. and Lengellé R. (1993), "Initializing back propagation networks with prototypes," *Neural Networks*, vol. 6, pp.351-363.

Devijver, P.A. and Kittler, J. (1982). Pattern recognition: A statistical approach. Englewood Cliffs: Prentice Hall.

Dodd, N. (1990), "Optimization of network structure using genetic techniques", Proc. Int. Joint Conf. on Neural Networks, San Diego.

Donoho, D. L. and Johnstone, I. M. (1989) Projection-based approximation and a duality with kernel methods, Annals of Statistics 17, pp. 58-106.

Drago G.P. and Ridella S. (1992), "Statistically controlled activation weight initialization (SCAWI)," *IEEE Trans. on Neural Networks*, vol. 3, no. 4, pp. 627-631, July.

Duda, R. O. and Hart, P. E. (1973) Pattern Classification and Scene Analysis. New York: Wiley.

Fahlman, S. E. and Lebiere C. (1990), "The Cascade-Correlation Learning Architecture", in Advances in Neural Information Processing Systems – 2, pp. 525-532, Morgan Kaufmann, San Mateo, CA.

Fogel, D.B. (1995), Evolutionary Computation: Toward a New Philosophy of Machine Intelligence, Piscataway, NJ: IEEE Press.

Frasconi, P., M. Gori and Tesi A., "Successes and Failure of Back-propagation: a Theoretical Investigation", chapter in Progress in Neural Networks, Ablex Publishing, Omid Omidvar (Ed.).

Fraser, A.M and Swinney, H.L. (1986). Independent coordinates for strange attractors from mutual information. Phys. Rev. A, 33(2), 1134-1140.

Friedman, J. H. and Tukey, J. W. (1981) Projection pursuit regression. Journal of the American Statistical Association 76, pp. 817-823.

Friedman, J. H. (1989) Regularized discriminant analysis, Journal of the American Statistical Association 84, pp. 165-175.

Fritzke, B. (1994). Growing cell structure: A self-organizing network for supervised and un-supervised learning. Neural Networks, 7(9), 1441-1460.

Fritzke, B. (1995). A growing neural gas network learns topologies. In G. Tesauro, D. S. Touretzky & T. K. Leen (Eds.), Advances in Neural Information Processing Systems 7 (pp. 625-632). Cambridge, MA: MIT Press.

Frohlinghaus, T., Weichert A. and Rujan P. (1994), "Hierarchical neural networks for time-series and control," Network, Vol. 5, pp. 101-106.

Funahashi, K. (1989), "On the Approximate Realization of Continuous Mappings by Neural Networks", Neural Networks, vol. 2, pp. 183-192.

Geman, S., Bienenstock E., and Doursat R. (1992), "Neural Networks and the Bias/Variance Dilemma", Neural Computation, vol. 4, pp. 1-58.

Goldberg, D.E. (1985), "Genetic algorithms and rule learning in dynamic system control", in Proc. 1st Int. Conf. On Genetic Algorithms and Their Applications, Hillsdale, NJ: Lawrence Erlbaum, pp. 8-15.

Golub, G.H. and Van Loan C. F. (1989), *Matrix Computations*, The Johns Hopkins University Press.

Gori, M and Tesi A. (1990), "Some examples of local minima during learning with Back-propagation", Parallel Architecture and Neural Networks, Vietri sul Mare (IT).

Gori M. and Tesi A. (1992), "On the problem of local minima in backpropagation", IEEE Transaction on Pattern Analysis and Machine Intelligence, vol. 14, no. 1, pp. 76-86.

Graepel, T., Burger, M., and Obermayer, K. (1997). Phase transitions in stochastic self-organizing maps. Physical Review E, 56, 3876-3890.

Graepel, T., Burger, M., and Obermayer, K. (1998). Self-organizing maps: generalizations and new optimization techniques. Neurocomputing, 21, 173-190.

Gray, R. M. (1984). Vector quantization. IEEE Acoust., Speech, Signal Processing Mag., 1 (2), 4-29.

Green, P. J. and Silverman, B. W. (1994) Non-parametric Regression and Generalize Linear Models. A Roughness Penalty Approach. London: Chapman & Hall.

Grinold, R. C. (1969) Comment on "Pattern classification design by linear programming", IEEE Transactions on Computers 18, pp. 378-379.

Halkidi, M., and Vazirgiannis, M. (2002). Clustering validity assessment using multi representatives. Proceedings of the 2nd Hellenic Conference on Artificial Intelligence.

Hall, M. A. (1999). Correlation-based Feature Selection for Machine Learning. Ph.D. thesis, Department of Computer Science, Waikato University, New Zealand.

Han, J.W. and M. Kamber. (2001). Data mining: concepts and techniques. San Meteo, CA: Morgan Kaufmann Publishers.

Hasitie, T., Tibshirani, R. and Friedman, J. (2001). The elements of statistical learning: Date mining, inference, and prediction. New York: Springer.

Hassibi B., Stork D. G., and Wolff G. J. (1993), "Optimal Brain Surgeon and General Network Pruning", Proc. IEEE International Conference on Neural Networks, San Fransisco, vol. 1, pp. 293-299.

Hastie, T. J. and Tibshirani R. J. (1990) Generalized Additive Models. London: Chapman & Hall.

Haykin, S. (1994) Neural Networks, a Comprehensive Foundation, IEEE Computer Society Press, 1994.

Haykin, S. (1994), Neural Network: A Comprehensive Foundation, Macmillan, Englewood Clifs, N.J.

Hebb, D. O. (1949), The Organization of Behavior: A Neuropsychological Theory, Wiley, New York.

Herrero, J., Valencia, A., and Dopazo, J. (2001). A hierarchical unsupervised growing neural network for clustering gene expression patterns. Bioinformatics, 17 (2), 126-136.

Heskes, T. (1999). Energy functions for self-organizing maps. In E. Oja and S. Kaski, (Eds.), Kohonen Maps (pp. 303-315). Amsterdam: Elsevier.

Ho, Y.C. and Kashyap, R. L. (1965) An algorithm for linear inequalities and its applications, IEEE Transactions on Electronic Computers 14, pp. 683-688.

Hochreiter S. and Schmidhuber J. (1997), "Flat Minima", Neural Computation, Vol. 9, pp. 1-42.

Holland, J.H. (1975), Adaptation in Natural and Artificial Systems, Ann Arbor, MI: Univ. of Michigan Press.

Hornik, K., Stinchcombe M. and White H (1989) Multilayer feedforward networks are universal approximators. Neural Networks, 2, pp. 359-366.

Hornik, K., Stinchcombe M., While H. (1989), "Multilayer feedforward networks are universal approximators", Neural Networks, vol. 2, pp. 359-366.

Horst, R. and Pardalos P.M. (1995), Handbooks of Global Optimization, Kluwer Academic Publisher, The Netherlands.

Huang, D, and Chow, T. W. S, Efficiently searching the important input variables using Bayesian discriminant, IEEE Trans on Circuit and Systems, Part I, Vol. 52, No. 4, April 2005, pp. 785-793.

Huang, D, and Chow, T. W. S,. Enhancing Density Based Data Reduction Using Entropy, Neural Computation, Feb 2006, Vol. 18, No. 2, pp. 470-495.

Huntsberger, T., and Ajjimarangsee, P. (1989). Parallel self-organizing feature maps for unsupervised pattern recognition. Int. J. General systems, 16, 357-372.

Jang, J. S. R. (1993) ANFIS: Adaptive-Network-Based Fuzzy Inference System, IEEE Transactions on Systems, Man and Cybernetics, 23(3) pp. 665-687.

John, G. H., Kohavi, R., and Pfleger, K. (1994). Irrelevant features and the subset selection problem. In Processings of the 11th international conference, eds. W. W. Cohen & H. Hirsh, 121-129, San Francisco, C. A : Morgan Kaufmann Publishers.

Joliffe, I. T. (1986). Principal component analysis. New York: Springer-Verlag.

Kenney, J. F. and Keeping, E. S. "Histograms." §2.4 in Mathematics of Statistics, Pt. 1, 3rd ed. Princeton, NJ: Van Nostrand, pp. 25-26, 1962.

Kiang, M. Y. (2001). Extending the Kohonen self-organizing map networks for clustering analysis. Computational Statistics & Data Analysis, 38, 161-180.

Kirkpatrick, S.C.D., Jr. Gelatt and Vecchi M.P. (1983), "Optimization by simulated annealing", Science 220, vol. 4598, pp. 671-680.

Kiviluoto, K. and Oja, E. (1998). S-map: A network with a simple self-organization algorithm for generative topographic mappings. In M. Jordan, M. Kearns & S. Solla (Eds.), Advances in Neural Information Processing Systems 10 (pp. 549-555). Cambridge, Massachusetts: MIT Press,.

Kohonen, Kaski, T., and Lappalainen, S., H. (1997). Self-organized formation of various invariant-feature filters in the adaptive-subspace SOM. Neural Computation, 9(6), 1321-1344.

Kohonen, T. (1997) Self-organizing maps, Springer-Verlag, Berlin, Germany, 1997.

Kohonen, T. (1998). The self-organizing map. Neurocomputing, 21, 1-6.

Kohonen, T. (2001). Self-Organizing Maps. London: Springer.

Koikkalainen, P. (1995). Fast deterministic self-organizing maps. Proc. Int. Conf. Artificial Neural Networks, 63-68.

Kolmogorov, A.N. (1957), "On the representation of continuous functions of many variables by superposition of continuous functions of one variable and addition," [in Russian], *Dokl. Akad. Nauk USSR*, 114, pp.953-957.

Köng, A. (2000). Interactive visualization and analysis of hierarchical neural projections for data mining. IEEE Trans. Neural Networks, 11(3), 615-624.

Kononenko, I. (1994). Estimating attributes: analysis and extensions of RELIEF. In proceedings of the european conference on Machine learning, eds. F. Bergadano & L. De Raedt, 171 – 182. Berlin: Springer-Verlag.

Kraaijveld, M. Mao, A., J., and Jain, A. K. (1995). A nonlinear projection method based on the Kohonen's topology preserving maps. IEEE Trans. Neural Networks, 6(3), 548-559.

Krogh, A. and Hertz J. A. (1992), "A Simple Weight Decay Can Improve Generalization", in S. J. Hanson, J. E. Moody and R. P. Lippmann, editors, Advances in Neural Information Processing Systems – 4, pp. 950-957, Morgan Kaufmann, San Mateo, CA.

Kryzak, A. and Linder T., (1998) Radial basis function networks and complexity regularization in function learning. IEEE Transactions on Neural Networks, 9(2), pp. 247-256.

Kubat, M. (1998) Decision trees can initialize radial-basis function networks, IEEE Trans. On Neural Networks, 9, pp. 813-821.

Kudo, M., and Sklansky, J. (2000). Comparison of algorithms that select features for pattern classifiers. Pattern Recognition, 33, 25-41.

Laakso, A., and Cottrell, G. W. (1998). How can I know what you think?: Assessing representational similarity in neural systems. Proc. 20th Annu. Cognitive Sci. Conf., 591-596.

Lampinen, J., and Oja, E. (1992). Clustering properties of hierarchical self-organizing maps. J. Math. Imag. Vis., 2(2-3), 261-272.

Lang K.J. and Witbrock M.J. (1988), "Learning to tell two spiral apart", in Proc. Of the 1988 Connectionist Models Summer School, Morgan Kaufmann.

LeCun, Y., Denker J. S., and Solla S. A. (1990), "Optimal Brain Damage", Advances in Neural Information Processing Systems II (Denver 1989), pp. 508-605.

Leung, C. T. (1998), Generalization Enhancement Methods for Neural Networks Under Finite Observed Data – Time Series Forecasting, PhD Thesis, City University of Hong Kong.

Leung, C.T. and Chow T.W.S. (1997), "A Novel Robust Fourth-order Cumulants Cost Function", Neurocomputing, vol. 16, Iss. 2, pp. 139-147, Aug.

Li, W.T. (1990). Mutual information functions versus correlation functions. Stat. Phys., vol. 60, no.5/6, 823-837.

Liu, H. and Motoda, H. (1998 b). Feature selection for knowledge discovery and data mining. London, GB: Kluwer Academic Publishers.

Liu, H., Motoda, H., and Dash, M. (1998 a). A monotonic measure for optimal feature selection. In Proceedings of European Conference on Machine Learning, 101-106.

Lowe, D., (1989) Adaptive radial basis function nonlinearities, and the problem of generalization, 1st IEE International Conference on Artificial Neural Networks, pp. 460-471, San Diego, CA.

Luttrel, S. (1994). A bayesian analysis of self-organizing maps. Neural Computation, 6, 767-794.

MacKay, D. (1992). A practical Bayesian framework for backpropagation networks, Neural computation, 4, 448-472

Mao, J., and Jain, A. K. (1996). A self-organizing network for hyperellipsoidal clustering (HEC). IEEE Trans. on Neural Networks, 7(1), 16-29.

Marsland, S., Shapiro, J., and Nehmzow, U. (2002). A self-organizing network that grows when required. Neural Networks, 15 (8-9), 1041-1058.

Meilhac, C,. and C. Nastar, "Relevance feedback and category searching in image databases," in Proc. IEEE Intl. Conf. Multimedia Computing and Systems, pp. 512-517, Florence, Italy, June, 1999.

Mendal J. M. (1991), "Tutorial on Higher-Order Statistics (Spectra) in Signal Processing and System Theory: Theoretical Results and Some Applications", Proc. IEEE, Vol. 79, No. 3, pp. 278-305.

Minnick, R. C. (1961) Linear-input logic. IRE Transactions on Electonic Computers 10, pp. 6-16.

Minsky, M. L. and Papert, S. A. (1988) Perceptrons. An Introduction to Computational Geometry. Expanded edition. Cambridge, MA: The MIT Press.

Mitchell, T. (1997) Machine Learning. McGraw-Hill.

Mitra, P, Murthy, C.A. and Pal, S.K. (2002). Unsupervised feature selection using feature similarity. IEEE Trans on PAMI, vol 24, no 3, 301-312.

Molina, L.C., Belanche, L. and Nebot, A. (2002). Feature selection algorithms: a survey and experimental evaluation. available at: http://www.lsi.upc.es/dept/techreps/html/R02-62.html. Technical Report.

Moody, J. and Darken C. (1988) Learning with localized receptive fields. In T. Sejnowski, D. Touretzky, and G. Hinton, editors, Connectionist Models Summer School, Carnegie Mellon University.

Moon, Y., B. Rajagopalan, U. Lall, "Estimation of mutual information using kernel density estimators," Phys. Rev. E, vol. 52, no. 3 (B), pp. 2318-2321, 1995.

Moon, Y., Rajagopalan, B., and Lall, U. (1995). Estimation of mutual information using kernel density estimators. Phys. Rev. E, 52(3), 3, 2318-2321.

Mueller, K. R., Mika S., Ratch G., Tsuda K., and Scholkopf B. (2001) An introduction to kernel-based learning algorithms. IEEE Transactions on Neural Networks, 12(2), pp. 181-201.

Mulier, F., and Cherkassky, V. (1995). Self-organization as an iterative kernel smoothing process. Neural computation, 7, 1165-1177.

Muneesawang, P., and Guan, L. (2002). Automatic machine interactions for content-based image retrieval using a self-organizing tree map architecture. IEEE Trans. Neural Networks, 13 (4), 821-834.

Muroga, S., Toda, I. and Takasu, S. (1961) Theory of majority decision elements. Journal of the Franklin Institute 271, pp. 376-418.

Murtagh, F. (1995). Interpreting the Kohonen self-organizing feature map using contiguity-constrained clustering. Pattern Recognition Letters, 16, 399-408.

Narendra, P.M. and Fukunaga, K. (1977). A branch and bound algorithm for feature selection. IEEE. Trans on computers, C-26(9), 917-922.

Neal, R., and Hinton, G. (1998). A new view of the EM algorithm that justifies incremental, sparse, and other variants. In M. Jordan (Ed.), Learning in Graphical Models (pp. 355-368). Dordrecht : Kluwer Academic.

Nguyen, D. and Widrow B. (1990), "Improving the learning speed of 2-layer neural networks by choosing initial values of the adaptive weights," *International Joint Conf. on Neural Networks*, vol. 3, pp.21-26, San Diego, CA, U.S.A., 17-21 July.

Nikias, C. L. and Petropuou A. P. (1993), Higher-Order Spectra Analysis: a Nonlinear Signal Processing Framework, Prentice-Hall, Englewood Cliffs, N.J.

Nikias, C. L., and Raghuveer M. R. (1987), "Bispectrum Estimation: A Digital Signal Processing Framework," Proc. IEEE, Vol. 75, No. 7, pp. 869-891.

Nojun, Kwak, and Choi C-H., (2002). Input feature selection for classification problems. IEEE. Trans on Neural networks, vol 13 no 1, 143-159.

Osowski S. (1993) "New approach to selection of initial values of weights in neural function approximation," *Electronics Letters*, vol. 29, no. 3, pp.313-315, 4th Feb.

Pal, N. R., and Eluri, V. K. (1998). Two efficient connectionist schemes for structure preserving dimensionality reduction. IEEE Trans. Neural networks, 9(6), 1142-1154.

Pal, N. R., Bezdek, J. C., and Tsao, E. C.-K. (1993). Generalized clustering networks and Kohonen's self-organizing scheme. IEEE Trans. Neural Networks, 4(4), 549-557.

Papoulis, A. (1991), Probability, Random Variables, and Stochastic Processes, 3rd international edition, McGraw-Hill, Singapore.

Park, J. and Sandberg I. W. (1991) Universal approximation using radial-basis functions. Neural Computation, 3, pp. 246-257.

Park, J. and Sandberg I. W. (1993) Approximation and radial-basis-function networks. Neural Computation, 3, pp. 305-316.

Park, J. and Sandberg I. W. (1991), "Universal Approximation Using Radial-Basis-Function Networks", Neural Computation, vol. 3, pp. 246-257.

Parzen, E. (1962). On the estimation of a probability density function and mode. Ann. Math. Statist. Vol 33, 1064-1076.

Pedrya, W (1998) Conditional fuzzy clustering in the design of radial basis function neural networks, IEEE Trans. On Neural Networks, 9, pp. 601-612.

Poggio, T. and Girosi F. (1990a) Networks for approximation and learning. Proceedings of the IEEE, 78 (9), pp. 1481-1497.

Poggio, T. and Girosi F. (1990) Regularization algorithms for learning that are equivalent to multilayer networks, Science 247, pp. 978-982.

Principe, J.C., Fisher III, and Xu J.D. (2000). Information theoretic learning. In Unsupervised adaptive filtering, eds. S.Haykin, New York, NY: Wiley.

Rao, C. R. (1985), Linear Statistical Inference and its Applications, 2nd Ed., Wiley Eastern, New Delhi.

Rauber, A., Merkl, D., and Dittenbach, M. (2002). The growing hierarchical self-organizing map: exploratory analysis of high-dimensional data. IEEE Trans. Neural Networks, 13(6), 1331-1341.

Ritter, H., Martinetz, T., and Schulten, K. (1992). Neural Computation and Self-Organizing Maps: An introduction. Reading, Mass: Addison-Wesley.

Rosenblatt, F. (1957) The perceptron – a perceiving and recognizing automation. Report 85-460-1. Cornell Aeronautical Laboratory.

Rosenblatt, F. (1958) The perceptron: a probabilistic model for information storage and organization in the brain. Psychological Review 65, pp. 386-408. Reprinted in Shavlik & Dietterich (1990).

Rosenblatt, F. (1962) Principles of Neurodynamics, Washington, DC: Spartan Books.

Rumelhart, D.E., Hinton G. E. and Williams R. J. (1986), "Learning representations by back-propagation errors", Nature, vol. 323, pp. 533-536.

Sadler, B. M., and Giannakis G. B. (1994), "Estimation and Detection in NonGaussian Noise Using Higher Order Statistics", IEEE Trans. Signal Processing, Vol. 42, No. 10, pp. 2729-2741.

Samet, H. (1984). The quadtree and related hierarchical data structures. ACM Computing Surveys, 16(2), 187-260.

Sammon, J. W. (1969). A nonlinear mapping for data structure analysis. IEEE Trans. Computers, 18(5), 491-509.

Sato, M., and Ishii, S. (2000). On-line EM algorithm for the normalized gaussian network. Neural Computation, 12, 407-432.

Schölkopf, B., and Smola, A. J. (2002). Learning with Kernels: Support Vector Machines, Regularization, Optimization and Beyond. Cambridge, Massachusetts: MIT Press.

Schölkopf, B., Williamson, R. C., Smola, A. J., Shawe-Taylor, J., and Platt, J. (2000). Support vector method for novelty detection. In Sara A. Solla, Todd K. Leen & K. R. Muller (Eds.), Advances in Neural Information Processing Systems 12 (pp. 582-588). Cambridge, Massachusetts: MIT Press.

Seiffert, U., and Jain, L. C. (2002). Self-organizing neural networks: recent advances and applications. New York: Springer-Verlag.

Setiono R. (1997), "A Penalty-Function Approach for Pruning Feedforward Network Networks," Neural Computation, Vol. 9, pp. 185-204.

Setiono, R., and Liu, H. (1997). Neural-network feature selector. IEEE Trans on NN, vol.8 no. 3, 654-661.

Shang, Y. and Wah B.W. (1996), "Global Optimization for Neural Networks Training", IEEE Computer, vol. 29, no. 3, pp. 45-54.

Shang, Y. and Wah B. W. (1996), "Global Optimization for Neural Network Training", IEEE Computer, Vol. 29, No. 3, pp. 45-54, Mar.

Shannon, C. E, A mathematical theory of communication, Bell System Technical Journal, vol. 27, pp. 379-423 and 623-656, July and October, 1948.

Shepanski, J.F. (1988), "Fast learning in artificial neural systems: multilayer perceptron training using optimal estimation," IEEE International Conf. on Neural Networks, vol. 1, pp.465-472, New York, IEEE Press.

Shimodaira, H. (1994), "A weight value initialization method for improving learning performance of the back propagation algorithm in neural networks," Proc. 6th Int. Conf. on Tools with Artificial Intelligence, pp.672-675, New Orleans, LA, USA, 6-9 Nov.

Si, J., Lin, S., Vuong, M. -A. (2000). Dynamic topology representing networks. Neural Networks, 13 (6), 617-627.

Silva F.M. and Almeida L.B. (1990), "Acceleration techniques for the backpropagation algorithm," *Lectures Notes in Computer Science*, vol.412, Springer-Verlag.

Silverman, B.W. (1986). Density estimation for statistics and data analysis. London: Chapman-Hall.

Smith, F.W., (1968) Pattern classifier design by linear programming. IEEE Transactions on Computers 18, pp. 548-551.

Su, M. C., and Chang, H. T. (2000). A new model of self-organizing neural networks and its application in data projection. IEEE Trans. Neural Networks, 12(1), 153-158.

Sutton, R.S. (1986), "Two problems with backpropagation and other steepest descent learning procedures for networks," *Proceedings of the 8th Annal Conf. of the Cognitive Science Society*, pp. 823-831.

Tax, D. M. J., and Duin, R. P. W. (1999). Support vector domain description. Pattern recognition letters, 20, 1991-1999.

Torn, A. and Zillinskas A. (1987), Global Optimization, Springler-Verlag.

Towell, G. and Shavlik J. W. (1994) Knowledge based artificial neural networks. Artificial Intelligence, 70(4), pp. 119-166.

Tresp, V., Hollatz J. and Ahmad S. (1993) Representing probabilistic rules with networks of Gaussian basis functions. Machine Learning, 27, pp. 173-200.

Tugnait, J. K. (1990), "Consisten Parameter Estimation for Non-causal Autoregressive Models via Higher-order Statistics," International Federation of Automatic Control, Vol.26, No. 1, pp. 51-61.

Ultsch, A., and Siemon, H. P. (1990). Kohonen's self organizing feature maps for exploratory data analysis. Proc. Int. Neural Network Conf, 305-308.

Unnikrishnan, K.P. and Venugopal K.P. (1994), Alopex: a correlation-based learning algorithm for feedforward and recurrent neural networks, *Neural Computations*, 6: 469-490.

Vapnik, V. N. (1995) The nature of statistical learning theory, Springer-Verlag, Berlin.

Vapnik, V. (1995). The nature of statistical learning theory. New York : Springer.

Vesanto, J., and Alhonierni, E. (2000). Clustering of the self-organizing map. IEEE Trans. Neural Networks, 11(3), 586-600.

Wang, H., Bell, D., and Murtagh, F. (1999). Axiomatic approach to feature subset selection based on relevance, IEEE Trans on PAMI, vol.21, no. 3, 271-277.

Wang, W., Jones, P. and Patridge, D. (2001). A comparative study of feature-salience ranking techniques, Neural Computation, 13(7), 1603-1623.

Weigend, A., Rumelhart D. and Huberman B. (1991), "Generalization by Weight Elimination with Application to Forecasting", Advances in Neural Information Processing III, R. P. Lippman and J. Moody Ed., Morgan Kaufmann, San Mateo, CA, pp. 875-882.

Widrow, B. and Hoff, Jr. M. E. (1960), "Adaptive switching circuits", IRE WESCON Convention Record, pp. 96-104.

Widrow, B and Hoff, M. E. Jr. (1960) Adaptive switching circuits. IRE WESCON Convention Record 4, pp. 96-104. Reprinted in Anderson & Rosenfeld (1988).

Wu, Sitao, and Chow, T. W. S. (2003b). Support vector visualization and clustering using self-organizing map and support vector one-class classification. International Joint Conference on Neural Networks (IJCNN), 803-808.

Wu, Sitao, and Chow, T. W. S. (2005a). Content-based image retrieval using growing hierarchical self-organizing quadtree map. Pattern Recognition, 38(5), 707-722.

Wu, Sitao, and Chow, T. W. S. (2005b). PRSOM - a probabilistic regularized self-organizing map for data projection and visualization. IEEE trans. Neural Networks, 16(6), 1362-1380.

Wu, Sitao, and T. W. S. Chow. (2004) Clustering of the self-organizing map using a clustering validity index based on inter-cluster and intra-cluster density. Pattern Recognition, 37(2), 175-188.

Wu, Sitao, Chow, T. W. S, Induction machine fault detection using SOM based RBF neural networks, IEEE Trans On Industrial Electronics, Vol. 51, No. 1, Feb 2004, pp. 183-194.

Xiao-Dong LI, John K. L. Ho, T. W. S. Chow, Iterative Learning Control for Linear Time-Variant Discrete Systems Based on 2-D System Theory, IEE Proceedings IEE Proceedings Control Theory and Applications, Vol. 152, No. 1, Jan 2005, pp.13-18.

Xu, L., Yan, P. and Chang, T. (1994) Best first strategy for feature selection. In Ninth International Conference on Pattern Recognition, 706—708, IEEE Computer Society Press.

Yam, Y.F. and Chow T.W.S. (1993), "Extended backpropagation algorithm," *Electronic Letters*, vol.29, no.19, pp.1701-1702.

Yam, Y.F. and Chow T.W.S. (1997), "A new method in determining the initial weights of feedforward neural networks," *Neurocomputing*, vol. 16, pp. 23-32.

Yam, J. Y. F. and Chow T. W. S., (2001) Feedforward networks training speed enhancement by optimal initialization of the synaptic coefficients, IEEE Trans. On Neural Networks, 2(2) pp. 430-434.

Yang, H. H. and Moody, J. (1999). Feature selection based on joint mutual information. In Advances in intelligent data analysis (AIDA) and computational intelligent methods and application (CIMA), Rochestor, New York.

Yin, H. (2002a). ViSOM-A novel method for multivariate data projection and structure visualization. IEEE Trans. Neural Networks, 13(1), 237-243.

Yin, H. (2002b). Data Visualization and manifold mapping using the ViSOM. Neural networks, 15, 1005-1016.

Zadeh, L A. (1965) Fuzzy Sets. Information Control, 8, pp. 338-353.

Zhang, Q. and Benveniste A. (1992) Wavelet networks, IEEE Transactions on Neural Networks 3, pp. 889-898.

Zhang, X., and Li, Y. (1993). Self-organizing map as a new method for clustering and data analysis. Proc. Int. Joint Conf. Neural Networks, 2448-2451.

Zhao, Y. and Atkeson, C. G. (1992) Some approximation properties of projection pursuit learning networks. In NIPS4, pp. 936-943.

Zheng Q. and Zhuang D. (1992), "Integral Global Optimization of Constraint Problems in Functional Spaces with Discontinuous Penalty Functions" Recent Advances in Global Optimization, edited by C.A. Floudas and P.M. Paradalos, Princeton University Press, pp. 298-320.

Index